habnf W9-AST-151 HEB
658.5752 DIANA

Diana, Carla, author
My robot gets me
33410017100746 04-15-2021

DISCARD

DISCARD

My Robot Gets Me

My Robot Gets Me

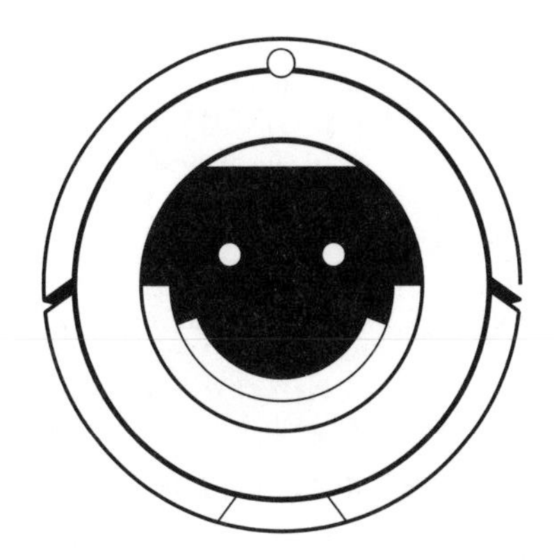

How Social Design Can Make New Products More Human

CARLA DIANA

Harvard Business Review Press
Boston, MA

HBR Press Quantity Sales Discounts

Harvard Business Review Press titles are available at significant quantity discounts when purchased in bulk for client gifts, sales promotions, and premiums. Special editions, including books with corporate logos, customized covers, and letters from the company or CEO printed in the front matter, as well as excerpts of existing books, can also be created in large quantities for special needs.

For details and discount information for both print and ebook formats, contact booksales@harvardbusiness.org, tel. 800-988-0886, or www.hbr.org/bulksales.

Copyright 2021 Carla Diana
All rights reserved
Printed in the United States of America

10 9 8 7 6 5 4 3 2 1

No part of this publication may be reproduced, stored in or introduced into a retrieval system, or transmitted, in any form, or by any means (electronic, mechanical, photocopying, recording, or otherwise), without the prior permission of the publisher. Requests for permission should be directed to permissions@harvardbusiness.org or mailed to Permissions, Harvard Business School Publishing, 60 Harvard Way, Boston, MA 02163.

The web addresses referenced in this book were live and correct at the time of the book's publication but may be subject to change.

Library of Congress Cataloging-in-Publication Data

Names: Diana, Carla, author.
Title: My robot gets me : how social design can make new products more human / Carla Diana.
Description: Boston, MA : Harvard Business Review Press, [2021] | Includes index. |
Identifiers: LCCN 2020047850 (print) | LCCN 2020047851 (ebook) | ISBN 9781633694422 (hardcover) | ISBN 9781633694439 (ebook)
Subjects: LCSH: Home automation. | Product design—Social aspects. | Artificial intelligence. | Human behavior. | Design and technology.
Classification: LCC TK7881.25 .D53 2021 (print) | LCC TK7881.25 (ebook) | DDC 658.5/752—dc23
LC record available at https://lccn.loc.gov/2020047850
LC ebook record available at https://lccn.loc.gov/2020047851

ISBN: 978-1-63369-442-2
eISBN: 978-1-63369-443-9

The paper used in this publication meets the requirements of the American National Standard for Permanence of Paper for Publications and Documents in Libraries and Archives Z39.48-1992.

For my darling boy Massimo.

You get me, always.

CONTENTS

My Robot Gets Me

1

Introduction

"Smart" Is Not Enough—Products Need to Be Social

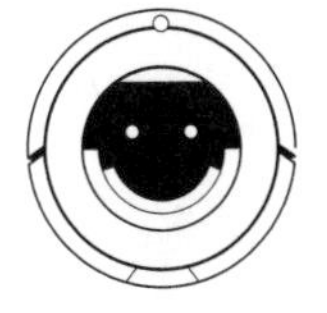

You're not crazy to thank Alexa when she checks the weather. It's not nuts to name your car *Keith* (or *Fred* or *Celeste*). And it's perfectly normal to want to repair your Roomba out of loyalty rather than replace it when it breaks.

An often-cited Georgia Tech study of people first using the Roomba Robot Vacuum—a device that looks like a twelve-inch round puck, with no deliberate features to make it seem like it has a face, body, or limbs—revealed that people saw it as a social entity, giving it names and talking to it directly.[1] I often thought much of this behavior stemmed from the gee-whiz factor or the novelty of having a "robotic" vacuum cleaner, but as a mainstream product that's been around since 2002 it still evokes this

FIGURE 1-1
The Roomba Robot Vacuum

kind of response. Here's a quick look at recent amazon.com reviews:[2]

> **Stephanie, "I can walk barefoot!!! Clean floors!,"** August 16, 2018. Roomba 675: "Our family immediately named her Hazel, and after several hours of hard work, I have no qualms about keeping her on my payroll. ;-)
>
> After Hazel ventured under the china cabinet (who knows when that area last saw my broom), my son said, 'Hey! The fuzzies are gone!! She's good!'"

> **Amya, "Time Saver/Allergen Reducer,"** December 2, 2018. "I love this little guy. We named him Wall-E because how could you not?
>
> The first few times he runs it will seem like a struggle (at least to me it did). But after about the third or fourth time, he seemed to be able to make his way around the

apartment with ease. He has gotten stuck a few times but has managed to wiggle free on his own."

Stephen and Susan, "Better than my husband," April 14, 2020. "It's definitely more reliable than my husband. It also does what I want . . . better than my husband. I turn it on and leave the house and when I come back . . . it at least did what I wanted it to. No excuses."

These examples are recent, but people have interacted socially with products long before there were computers, LEDs, or microchips. We have embraced special pillows, caressed key holders, and praised our washing machines. The social connection between person and product is only amplified when we add *smartness*, or the ability of a product to sense what we're doing and respond appropriately. Behaviors like these are informed by the social norms, psychological responses, and interaction patterns that we have with other people, even if the things we're interacting with are not people and we know it. Moreover, these tendencies grow stronger rather than weaker the more people interact, and great design plays a big role in the success of the interaction. The better the social exchange, the deeper the relationship a person has with a product and the better the experience overall. As technology becomes more sophisticated and microprocessors shrink to become embedded in everything from tennis rackets to syringes, we can program their behaviors as though they were little robotic entities.

As humans we are driven to be social. We have evolved to benefit from, value, and enjoy living together in groups, and we instinctively seek out interactions with people around us, something the 2020 pandemic has made us acutely aware of. The social scientist Michael Argyle argued that people are social animals for evolutionary reasons—that is, community structures help ensure the food and shelter we need to survive.[3] This makes us so highly attuned to social signals that we perceive them even when they aren't there: we see faces on products when there is no humanlike or animallike anatomy in the design and attribute emotions and intent to inanimate objects.[4] We talk about a broken product as being "sick" or "on strike." People will often call into customer service desks complaining that their devices "don't like them" or are "acting up."

Objects can't serve as substitutes for that human closeness we all crave; however, our deep need for social interaction makes us prone to interpreting interactions with products as social whether we are aware of it or not. Even devices that appear to behave in only vaguely communicative ways evoke sociality-based reactions.

Product Psychosis: Why "Smart" Products Aren't So Smart

As the consumer products we own start to sense things in our environment, move about our living and working spaces, and perform tasks on our behalf, the importance of the social interactions we have with devices will only grow. Some people think the instinct to interact socially with products is something to fight against, or a sentimental weakness to overcome. To designers like myself who are immersed in the practice of creating meaningful interactions, this is like trying to fight people's preferences for a variety of colors or their need to have written materials laid out in a manner that is easy to read—the best products come from working with people's preferences, tendencies, and limitations. Our designs can work in harmony with natural tendencies, and to work well, social qualities should be *designed on purpose*, rather than left to happen by accident. This book makes explicit something that successful interaction designers—who are also essentially "social designers"—do as an inherent and integral part of their work.

As technology's influence on our lives continues to rise, Hollywood has us fearing the specter of being taken over by intelligent machines, but the scary thing is that we have already been taken over by machines that aren't really intelligent at all. While it may seem that concern about robotics and artificial intelligence (AI) prevails in popular media, people are nonetheless eagerly purchasing consumer products with increasingly "smart" features. Many so-called smart products on the market are like the cringeworthy cousin who comes to visit and with whom you find it difficult to relate no matter how hard you try. After a day of awkward exchanges, you can't wait for him to leave. Between his perpetual interruptions with left-field anecdotes when you're concentrating on cooking and his inappropriately regaling your five-year-old with off-color jokes, you need a moment in a scream room just to recover.

How Do Today's Products Fall Short?

To answer this, we'll take a look at the automatic door, a product so ubiquitous and commonplace that we seldom question why it behaves so poorly from a social point of view. Dr. Wendy Ju, assistant professor at the Jacobs Technion-Cornell Institute at Cornell Tech and in the information science field at Cornell University, describes this interaction vividly in her book, *The Design of Implicit Interactions*.[5]

Imagine a doorman who does not acknowledge you when you approach. He stands there, silent and still, until you inch your way closer to the entrance. When you get to a distance of about two feet, *boom*! He abruptly flings the doors wide open. If you arrive after hours, you may find yourself waiting uncomfortably before you come to the realization that the doors are locked. The doorman's blank stare, unfortunately, gave you no clue. Upon meeting such a gatekeeper, many of us might run away, suspecting psychosis, yet this discomforting behavior is typical of our day-to-day interactions with not only automatic doors but many other devices designed to improve the quality of our lives. From Amazon's personal assistant Alexa, which eavesdrops without reminding us that it's there, to automatic lights that cloak us in darkness if we sit too still in a desk chair, many of the products we rely on lack sensitivity to the social cues and norms humans use to communicate and interact with one another.

Despite their failures as social entities, responsive products are here to stay and increasingly becoming essential in our cooking, cleaning, entertainment, health care, security, and hygiene. We are at a watershed moment in which technology can support almost any function that product designers and entrepreneurs envision, and the opportunities are limited only by our imagination. When a product does achieve the nirvana of authentic human interaction, studies show that its creators are rewarded with greater customer engagement, satisfaction, and loyalty. Despite this success, too many products still rely on crude behavioral designs that flunk the "normal" test. The solution, therefore, is not more technology but better—that is, more human and social—design.

Integrating Social Design into Product Creation

This book focuses on enabling technology to humanize consumer products through socially informed interactions. It will illuminate how sensitive people are to social, cultural, and personal norms and why it is important for products to adapt appropriately to these contexts. Through case studies, lab experiments, and interviews with experts it will showcase how design makes the difference between good and bad products, providing inspiration and advice to anyone who wants to embrace today's technology through new products. Each chapter will provide vivid examples of good and bad product design, along with concrete guidelines about how to conceptualize, build, and optimize interactive products. The aim is to cut through the intimidating chatter of high tech to showcase a vision for smart technology that understands and adopts the social values, norms, and protocols that we use to interact seamlessly with one another every day.

Social designers anticipate the exchanges people will have with a product in different situations and then craft the details of the interaction, taking into account how the product looks and feels throughout a typical exchange (and, ideally, throughout the life of the product). This will mean building a script of sorts that details what actions and reactions will take place, what the person's needs are, and how the product can fulfill those needs. Though that script may sometimes include verbal messages such as "Battery charging is complete" or "Service required soon," it just as often involves other modalities, such as light, sound, and motion, that communicate the same messages, only through abstraction, such as a flashing red light that indicates something is amiss.

Although the designers are used to thinking about the communicative abilities of formal characteristics such as color, typography, and materials, these more dynamic characteristics are a whole new ball of wax, giving an object the appearance of agency and making it feel "alive." This pioneer territory requires designers to work together with engineers, psychologists, programmers, design researchers, and marketers to acknowledge and collaboratively create a lexicon around how social characteristics are applied to product creation.

From a business perspective, it will be critical to build teams of people who can understand the technologies of interaction, as well as the psychological and social rules that govern people's responses to different interaction designs. The next generation of product designers and team leaders will need a broad suite of skills beyond those taught to designers today.

In other words, making sure that our products "get" us is a complex task that requires a great deal of insight, planning, and exploration. There's no one formula, and it's not all about algorithms. For the human touch to emerge as an outcome of product development, we need to focus on crafting the right social interaction from day one. That's what *My Robot Gets Me* is all about.

How Social Design Works—A Framework

Social interaction is complex and many layered. For an appropriate depth of understanding in creating socially savvy products, we must draw on insights from the social and behavioral sciences, from engineering and computer science, and from marketing and management science. The integration of these multiple perspectives is required to design the social lives of products

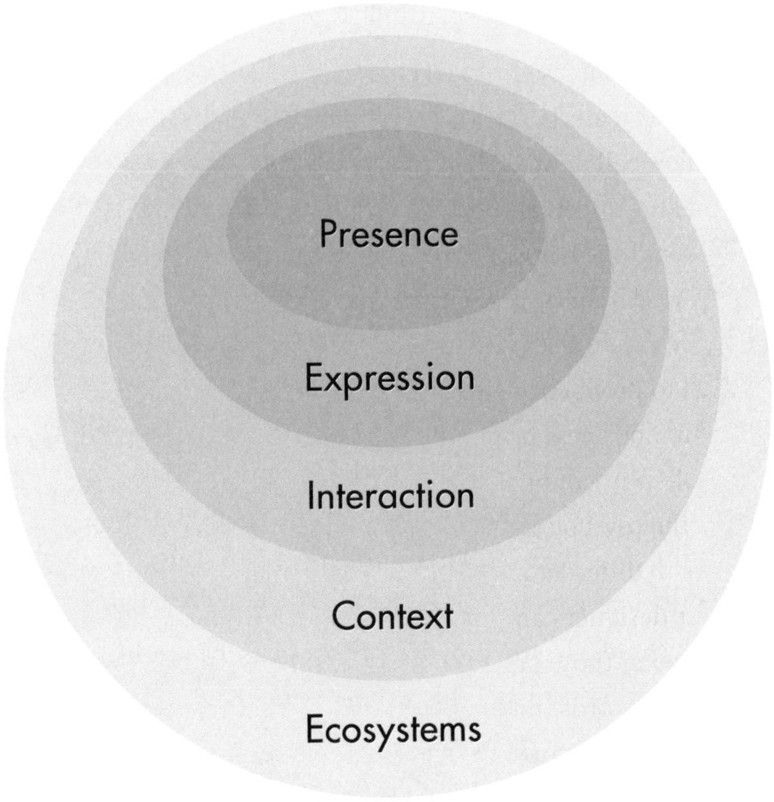

Presence

Expression

Interaction

Context

Ecosystems

FIGURE 1-2
The Five Rings in the Social Life of Products Framework

well. This may be why well-designed interactive products are still rare.

The social interaction we have with products should be considered and articulated by a product team before any other design decisions are made. When armed with the knowledge and techniques described in this book, product managers will hopefully encourage their teams to reflect more deeply on the nature of our relationships with products to envision great interactions before talking about features or technical characteristics.

My Robot Gets Me is organized using a framework that maps out the expanding scopes of interaction. It starts at the core, examining the *presence* of socially interactive products themselves—examples, core technologies, component parts—and goes on to discuss interaction between people and their products and the designers who make these products. Next is a look at *expression* in interactive products—what is communicated by interactive products, how, and to what effect. *Interaction* goes beyond expression to think of the back-and-forth dialogue that occurs when products can sense and respond to people. The types of interactions and what they mean depend a lot on the *context* in which they take place; this includes not only the environment in which interaction is occurring but also the task, timing, purpose, and role of the interaction. Encompassing everything is the *ecosystem*, which accounts for the broader product family, product ecosystem, and business model that influence the how and why of the product and its interactions.

As we move from focusing on the product toward the larger ecosystem that products are in, we're also able to move from the *what—robot vacuum cleaner!*—to the *how* and *why—keeping my house clean without having to get dirty*. All parts of the product are operating simultaneously and are critical to product success, but each part is complicated enough that it needs to be attended to separately.

By examining all the different layers of a product's social life, we can gain a big-picture view of all the factors that influence the interactions we will have with future products and how many levels of design need to be addressed to create successful social interaction with people. By focusing on scales of concern rather than discipline, we hope to bring together the disparate areas of study and expertise needed to address the design of interactive products at each level.

Social Product Literacy Is for Everyone

As technology-based products become more sophisticated, an understanding of their inner workings becomes increasingly difficult to decipher. Rather than have the average consumer believe that their interactive products operate by "magic," we need to spark and develop an interest in how and why products work. And this goes for everyone involved in product creation and management. While this book is aimed at professionals directly involved in the development of products for consumer use, it can also appeal to a general audience of product enthusiasts. An increased understanding of social product design will lead to a higher demand for socially well-designed products that are highly crafted and thoughtfully executed

At this moment in history, we're on the brink of an important new change in product design. No longer are we just making devices for the highly motivated 1 percent who are technically savvy and prepared to learn the ins and outs of complex systems, however tedious that might be. Now, the user base for interactive devices includes children and the elderly, people who have different levels of language proficiency and varying degrees of experience with electronic devices. A recent Pew Research Center report states that "Roughly two-thirds of those ages 65 and older go online and a record share now own smartphones."[6] Whereas a smart watch may have seemed like a frivolous and complex gadget suited only for early adopters, we now see people buying them for their Luddite family members, such as my friend Susan, who recently gave an Apple Watch to her ninety-year-old father to enjoy for its elegant styling and hands-free messaging. The fact that the watch will alert her should he happen to fall is a bonus that everyone appreciates regardless of what they think of the intrusion of "modern technology" into everyday life.

As interactive devices become more portable and robust, we are designing not only for work or home environments but for every environment people find themselves in, be it on the road, in a crowd, or even underwater.

The spring of 2020 brought with it a profound and abrupt change to daily life with the tragedy of the Covid-19 pandemic. People went from a Friday evening of hanging out in restaurants, attending parties, and strolling through galleries to a shut-in Monday morning of sheltering in place, working remotely, and leaving home only when necessary for food, medication, or life-threatening emergency needs. People scrambled to make up for the loss of social contact, turning to technology to fill the gaps. People who "hated" video chat were suddenly the family experts on Zoom's multiparticipant features, organizing virtual Trivial Pursuit games and online parties. Residents of nursing homes and assisted-living facilities were given crash courses in FaceTime and WhatsApp to keep in touch with loved ones. Going to work meant logging on from the living room, and *telemedicine* became a household word. And others who had never given robotics a second thought were waxing philosophical about the potential for automated robot delivery services, no-contact grocery pickup lockers, and roving disinfection devices.

Although the research and writing for this book were in progress for several years before the pandemic, the advent of Covid-19 highlighted the importance of this subject more than ever before. When people had to suddenly learn to use new devices to work, feed their families, and get medical advice, they needed to intuit how to communicate with the devices on hand, and pronto. There was no time for poring over user manuals or following elaborate tutorials, and the products with the stronger social abilities are the ones that won people's trust and loyalty because they just simply worked in the way people expected.

Products at the Intersection of the Physical and the Digital

While *product* can refer to a piece of software or an app, this book will look holistically at every aspect of the experience, taking into consideration the product's materials, forms, and relationship to the body as well as any digital characteristics, such as screen-based messaging, and the dynamic characteristics of light, sound, and movement. We'll consider obvious examples of social design, such as smart assistants, chatbots and voice agents like Siri, Alexa, Cortana, and Google Assistant, but will also look at those with more subtle interfaces that may not have an actual voice at all and respond with what might be thought of as a robotic language, like the chirps, lights, and motions of *Star Wars*' R2-D2 or Pixar's WALL-E. They might be as simple as a microphone that pivots toward the person speaking or a wristband that vibrates to communicate an incoming message.

The most powerful aspect of our relationships with our products will be the split-second, near-telepathic exchanges that can happen when we learn to decipher representational messages such as a flicker of light, a sequence of tones, or a gestural movement—the kinds of messages that can benefit from our full attention yet can also take place in our peripheral vision. Furthermore, products with sophisticated and contextually sensitive interaction will have learned from their exchanges with us. They'll maintain a memory of what we like, dislike, understand, or don't understand, and they will ultimately adapt to best serve us.

A product's ability to navigate physical spaces through light, sound, and motion give it the power to not only occupy an area but transform it as well, illuminating parts of a room, occupying walls, floors, or tables in varied ways, or beckoning with sounds from another room. Our relationship with these products will grow beyond the one-on-one, flat experience of a face in front of a screen to encompass a full-bodied interplay, a dance of sorts between a person and the objects in his or her environment.

Why Me?

I've always had a passion for creating objects. I began my career as a mechanical engineer, a role that seemed like the logical path for someone wanting to make things by combining creativity and a penchant for technical pursuits. After over seven years of combined experience as a design engineer and product researcher, I decided to reflect on my experience by returning to graduate school to explore the human side of product creation through a master's program at Cranbrook Academy of Art, an institution recognized for its rich history in American design—Eliel Saarinen and Charles Eames were founding faculty members—as well as being a hotbed of provocative idea generation. At Cranbrook it was thrilling to recognize that great design work consisted of more questions than answers. In my own research in the late 1990s, my inquiry focused on the potential for blending the physical and the digital in a holistic experience in which objects could come to life through programmed behaviors. I began with a series of experimental projects and haven't stopped exploring since.

After graduate school, I had the great fortune of working at some of the world's best design firms, including famed designer Karim Rashid's office and product consultancy frog design, eventually finding focus at the innovation design firm Smart Design, where I created the Smart Interaction Lab, an initiative focused on design explorations in the form of tinkering and hands-on experimentation around topics such as expressive objects, digital making, and presence and awareness. My work at Smart honed in on interactive physical products, and it's there that I began to recognize that the ideas about smart objects being explored by me and my colleagues in academia were beginning to find their way into everyday objects in a real and practical way. I led several design teams working on a range of products, from kitchen appliances to medical devices, toys, and automotive interiors, and served as a hybrid in a totally new territory of design that lay at the intersection of the physical and the digital.

My work in this new area brought me to a number of wonderful work experiences, and I found myself drawn to both industry and academia; I loved the satisfaction of directly influencing people's lives through the former and the thrill of placing one foot in the future through the bleeding-edge research of the latter. I have managed to find a balance between the two, continually teetering on the edge of different worlds. I've enjoyed sharing my passion by translating ideas from engineers to creatives and vice versa. The projects I've worked on have appeared on the covers of *Popular Science*, *Technology Review*, and the *New York Times Sunday Review*,[7] as well as *Time* magazine's top inventions of 2019, and my excitement for exploring the frontiers of product interaction has led to my being frequently called upon to write and speak internationally on the social impact of design and technology on our everyday lives.

Before joining Smart, I began a decade-long relationship as a key member of the Socially Intelligent Machines Lab founded by Dr. Andrea Thomaz at the Georgia Institute of Technology in 2007.[8] Together with a mechanical engineer named Jonathan Holmes, we were the core team for the lab's seminal social robot platform, named Simon, which was used to study all the ways we might interact with computing machines in an intuitive and human way by gesturing, speaking, exchanging tools and other objects, and working collaboratively in a variety of environments, such as kitchens and lobbies.[9] When Andrea brought her research and her lab to the University of Texas at Austin, we continued our collaboration, giving me the chance to dive deeper into social robot design while gathering learnings from human-robot interaction research that I could apply to other consumer product work taking place at Smart and, eventually, in my own independent studio. In 2017 Andrea founded Diligent Robotics, a company that creates robotic products for the health-care industry, and I had the honor of coming on board as the company's head of design. Diligent's signature product, Moxi, is a highly interactive robot that's currently deployed in a number of hospitals in the United States.

While not every product will be a complex robot such as Moxi, the nuances of interaction that take place on a regular basis between a robot and the people who use it can be applied to many types of products, even those as seemingly mundane as toothbrushes, coffee makers, microphones, and scooters. The potential for applied robotics to improve our social interaction with everyday objects is nothing short of remarkable and, as the

FIGURE 1-3
Moxi, the Highly Interactive Hospital Robot

inspiration for *My Robot Gets Me*, will inform stories and insights that appear throughout the book.

In addition to my work as a design consultant in the robotics realm, I have applied my passion to the creation of the 4D Design Department at my alma mater, Cranbrook.[10] A highly selective MFA program for students who want to explore the intersection of code, form, and electronics, it functions as an experimental laboratory for creative applications of technology. The department explores the myriad ways that the physical world around us has become infused with an undercurrent of flowing data, turning everyday experiences into connected, feedback-driven interactions that are transforming every aspect of culture and society. As the academy's first new program in forty-seven years, it builds on Cranbrook's historic legacy of experimental design activities while redefining craft to encompass a broad range of outcomes, including interactive objects, projected images, embedded electronics, applied robotics, computer-controlled machining, three-dimensional printing, and mixed-reality environments.

This book is based on material that I teach at Cranbrook 4D that has evolved from the courses I created for the School of Visual Arts, University of Pennsylvania, and Parsons School of Design, some of the country's first courses focused on designing smart objects. To complement my teaching and client work, I cohost the *RoboPsych* podcast, a biweekly discussion around design and the psychological impact of human-robot interaction.[11] Together with Tom Guarriello, a PhD psychologist, branding expert, and *RoboPsych*'s founder, we explore the impact of robotics on society and culture and look toward future products and systems through in-depth interviews with experts.

My Robot Gets Me blends the content I've created for my teaching, speaking, podcast hosting, and writing with the insights from my experience as a consumer product designer along with the ongoing learning from Diligent's cutting-edge robots that are currently in use out in the field. Throughout this text you will see commentary and wisdom from colleagues in industry and academia blended with anecdotes and interviews about projects that I've observed over two decades of pioneering work.

How to Use This Book and Who It's For

My Robot Gets Me is meant for a wide audience. Readers involved in the product development process but who do not have formal training in design can use it as a primer that ranges over contemporary ideas and practice in interaction design and user experience design. It will help them to better connect with colleagues who are deeper in the trenches of product development and working on creating the product's inner workings. People with a background in design will recognize familiar concepts in the first few chapters, with an expansion of fundamental design ideas to encompass the world of smart objects as the text progresses into the later chapters. It will be helpful as a way to gather deeper knowledge around applying robotics to product design; it can also serve as a useful text to share with less design-oriented colleagues throughout an organization in order to form a bridge for idea exchanges, allowing everyone to share a mental model of a product's potential and collaboratively envision opportunities for product and service creation. Ideally, it can help people to advocate for a social approach rather than a tech/features-based approach in their own organizations, supporting those championing the big picture in product interaction and getting everyone in the organization on board. Finally, it may also appeal to a general audience of readers who have an enthusiasm for contemporary product design and want to learn more about how things are created in order to better appreciate their uses and make smart purchasing decisions.

The Organization of the Book

My Robot Gets Me takes a layered approach to the idea of the product as its own social entity. Using the product context framework described above, the book will explain how to conceive of the critical aspects of an interactive device, taking into account its physical presence; the ways it expresses itself; techniques for seeing, hearing, and understanding people and its environment; its human social context; and its relationship to services and other products in an ecosystem. Each chapter relates to a ring in the framework and will explore how interactive devices function at that scope, presenting illustrative examples and analyzing key aspects. Each chapter will wrap up with guidelines and principles for how to design interactive devices that come with that framing of the product.

How Social Design Works: Affordances and Interaction

Chapter 2 offers a foundation for concepts that will be expanded upon throughout the course of the book. It will consider how key aspects of cognitive science set the stage for how a product is perceived by a person using it and provide the groundwork for product creators to envision how people and products communicate with one another.

Product Presence: Form Follows Feeling

Chapter 3, on presence, represents the core of the framework, delving into the overall impression of a product, beginning with its physical shape and material characteristics. It will discuss how physical attributes establish the foundation of a social relationship between people and their products and why the physical world still matters despite our devotion to the immaterial world of screens, apps, and software.

Object Expression: Communicating Behavior

Chapter 4, on expression, looks at how a product communicates outwardly to the people using it. It will look at core messages that a product needs to communicate, such as aspects of its operation and information that it needs to proceed in completing tasks. And it will explore how the basic modalities of sound, light, and motion can contribute to messaging that can be effective through both nonverbal and verbal cues.

Interaction Intelligence: The Rich Conversation between Objects and People

Chapter 5, on interaction, delves into the complexities of communication between product and person as they relate to the next ring in our framework. It starts with an understanding of how products can communicate messages outwardly, as explored in chapter 2, but then branches out to look at what happens when it's taking in and responding to sensor data, creating an ongoing and ever-changing feedback loop. It will provide an overview of how products understand the people and world around them and then explore key patterns for effective exchanges.

Designing Context: The Right Interaction for the Right Time and Frame of Mind

Chapter 6, on context, is about how the human social context in which a product will be used should affect every design decision. It starts with a look at the physical situation in which the product's use takes place (home, work, outdoors, indoors, winter, summer, etc.) and continues to include a consideration of the person's state of mind when using the product (calm, anxious, engaged, distracted, etc.). Good design will take into account context, using data input from the person and the environment to determine changes in context and respond appropriately.

Designing Ecosystems: Connecting Everything Together

Chapters 3–6 will look at individual products pretty much in isolation, but many powerful experiences happen when products interact with one another as part of a system, such as the way a smartphone, watch, and speaker might work together to offer calendar information and verbal reminders. Chapter 7, on ecosystems, looks at how services are linked to a host of different products that can respond individually depending on use cases and contexts while also acting in unison to provide data to the larger system.

Intelligence on Many Levels: AI and Social Savvy

Chapter 8 provides a look at where social objects are going next by gauging the success of today's smart products while also examining the technology trends at the forefront of academic and industry research that focus on social intelligence as the future of product intelligence. It examines how evolving capabilities will shape how we integrate products into our daily lives and what types of interactions we can hope to achieve.

The Future Is Here: Now What?

Chapter 9 offers a review of key concepts explored throughout the book and challenges designers to reflect upon how human values can be supported by new product relationships.

Each chapter makes a case for what matters at the scope of its layer, how it influences the social life of products, and what designers need to consider at that level. Join me as I embark on the journey of considering how social products are developed, starting with the fundamental importance of what an object's physical presence signifies to the people who will use it.

2

How Social Design Works

Affordances and Interaction

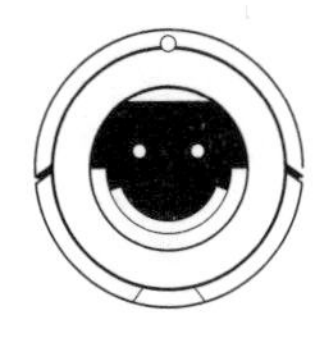

In thinking about the social design of smart objects, I often reflect on the relationship I have with my laptop, which could in some ways have been thought of as the ultimate, all-purpose smart object (that is, before the advent of the smartphone). As cold and hard as it might be, over time it has grown to something that feels natural. When it's closed, it's my accessory, a thick metal stand-in for a notebook, carried under my arm and reminding me of the potential for creativity and connection with other people. When open, it transforms into an entity that demands my attention. Its screen stands upright, directly facing me and blocking the view of anything beyond it. Glowing brighter than anything around it, I'm drawn into the activity happening on the illuminated rectangle. No longer a sidekick, it becomes the focus of my attention, supplanting any interaction that might be taking place and becoming the main conversation. I perform an action on the keyboard, which results in activity through light or sound. I am completely immersed, sometimes for hours, at the exclusion of everything and everyone else around me.

A short caress of the trackpad sends the page traveling upward, allowing me to review the outline originally written above this. Sometimes I absentmindedly use this little caress to reinforce the connection I have with the machine, nudging the pages up and down on the screen to bounce around along with my thoughts. Many times during the interaction, I am one with the machine, my hands flying around the keyboard, taking advantage of muscle memory that's been developed over decades and is so second nature that I'm not even cognizant of how the thought starts and how it ends.

This intuitive and intimate interaction is a good example of social interaction—one in which back-and-forth communication requires little conscious thought or mental gymnastics; however, this intimacy did not come easily. Undeniably, the way we control our computers—through mice, keyboards, and trackpads—is an uncomfortably unnatural way of interacting. (In my years of teaching design tools to art students at the graduate and undergraduate level, I have memories of several students who were new to computers assuming that the mouse should be placed right on the screen in order to control the cursor. Though this seems absurd to anyone who has trained themselves to use a mouse, it does fundamentally make sense.) It exists, at best, in what I would call the "awkward teenage-years" progression of computing devices—that is, it is good *enough* to serve its purposes and was the best manufacturers could do with the components and their respective costs at hand, but it is far from ideal.

Despite the challenge, I have developed an intimacy with my laptop, and this relationship is just one example of a social interaction that takes place in everyday life. Other social exchanges take place between me and my coffee machine, my mobile phone, my thermostat, and my car dashboard. Some take place through taps and caresses, and as microphone and camera technologies become more sophisticated and ubiquitous, others increasingly take place through voice and physical gestures, but they still add up to an enormous cognitive load perpetually foist upon me throughout the day.

In the past it was not always economically feasible to develop products with truly sophisticated and intuitive interfaces—sensor components were expensive, software was onerous to develop, and microcontrollers were not small and affordable enough to make sense in many consumer product applications. We knew computing devices were woefully inadequate, and we ardently envisioned, explored, and even prototyped ideal solutions, but it was rarely possible to implement these new ideas in mass-produced products. It was clear to us that experimentation around the potential for new ways of interacting would prove valuable to future product design efforts.

With so many technical hurdles now behind us, designers can lean on decades of ideal visions to inform new product solutions and bring them to life in real, concrete ideas and viable product proposals.

Social Affordances

This book will focus on ways to create and measure the social affordances of new products. *Social affordance* is a term that builds upon cognitive psychologist Donald Norman's definition of designed affordances, with the specific filter of social value applied.[1] It was originally used by Erin Bradner, research scientist at Autodesk, to consider how communications between people that are mediated by technology (such as email, messaging systems, and conferencing technologies) can facilitate group work.[2] Here, we expand upon the definition to include social interactions that take place between a person and their product. In other words, the social interaction can be the result of using the device as an avatar, or stand-in, for a person, or it can be the vehicle for communication between the person and the product as *an entity in itself.*

Social affordance for product design has a few aspects to consider:

- *Degree of intimacy:* Defining the nature of the relationship on the self-other continuum. Is this product an extension of myself, or is it an agent that acts on my behalf through tasks that it conducts independent of my direction?
- *Form:* How the physical shapes that make up the product's anatomy communicate social abilities and intentions and relate to the human body.
- *Dynamic behaviors:* How light, sound, and motion changes influence the social interactions between person and product.

- *Conversational elements:* How the messages conveyed by the product shape the way in which the relationship between product and person evolves.

Social Affordances and Better Product Experiences

Fortunately, we are in a new era in which those of us creating products can do much better, diminishing the learning curve for getting accustomed to an interface and thus building a stronger bond, making products easier to use and accessible to a greater portion of the population. Imagine my laptop as an extension of my body, allowing me to move the cursor where I want it just by shifting my gaze instead of having to manage the removed interface of the trackpad. Perhaps it could be a thin, flexible interface that I unfold so that it's in my pocket whenever I need it. And maybe its main interface can shift from being a keyboard to being a drawing surface so that it allows me multiple ways to input ideas.

In this way it could serve as a prosthetic, a natural part of myself. Another type of relationship one could have with a computing device would be as another entity with whom it would be natural to interact in social ways. As a designer, a big "aha" moment for me in terms of the value and power of social interaction with computers came in 2007, when I became part of the core team invited to develop a robot named Simon at the Georgia Institute of Technology in the Socially Intelligent Machines Lab led by Dr. Andrea Thomaz.

Simon was an upper-torso humanoid robot being developed to look at ways for people and machines to live and work alongside one another, and I was brought on board to lead the creative aspects of designing the robot's overall architecture, which would serve as a core means for defining its movement and behavioral characteristics. Together with Andrea and mechanical engineering partner Jonathan Holmes, we set out to create a robot that could explore how we might be able to control and train computers without needing any knowledge of code, mechanics, or button presses. The goal of the Simon project was to try to create a computing machine that relied only on social cues for its control and performance. In other words, to interact with the robot, one needed only to approach it with the skills he or she already had in interacting with people: talking, gesturing, exchanging objects, and so on. Simon understood spoken sentences and used voice, movement, and light behaviors to respond appropriately. If it didn't understand a certain request, it raised its arms in an apparent plea for forgiveness or cocked its head to express confusion. Its ears lit up when it recognized a color, and it spoke back when a person finished talking.

Simon was a one-off prototype developed to study people's interactions. Although I can describe how intuitive it is for me to use my laptop, it actually took a lot of cognitive effort at some point in my life to learn how to interact with it. Whether aware of it or not, I spent years training myself to use a QWERTY keyboard, originally as a teenager on an old-fashioned typewriter and eventually on a computer keyboard similar to the one that's part of the laptop case. The trackpad that feels so smooth under my fingertips was also a device I had to learn to use, gradually building an understanding of how to position my finger on the small rectangle while looking away at the larger rectangle in front of me onto which it was mapped.

FIGURE 2-1
Simon, the Robot That Relies on Social Cues

Simon's Ears

Simon was by far the most exciting project I had worked on up until that moment, and as a designer interested in interactivity, this was about as extreme an interface as one could get. But before we could get to thinking about how (and if) the robot might speak, move, or light up, I had to develop a clear direction on its overall shape and form. Relying on Andrea's research insights from previous robot projects, I knew it would need to have an expressive head that was capable of multiple gestures. It would need eyes that could indicate gaze (where and what it was looking at) and show social attention (with whom it was communicating). But one feature of the robot's anatomy that I may have taken for granted before this was what might seem to be functionally unnecessary: ears.

Ears serve no mechanical function for a robot—the microphones needed to understand speech can be well hidden within the head or torso. In fact, Simon's ears don't actually house the microphones that enable it to "hear." Instead, these appendages serve a purely social function, offering both logistical clues and emotional feedback. Andrea had initiated the Simon project after her PhD research on Leonardo, a seminal project in social robotics under the auspices of Dr. Cynthia Breazeal.[3] The team that had worked on Leonardo learned that the robot's ears provided people with important nonverbal information regarding what was happening during their interaction with it, much the way that people decipher the meaning of ear movement on a cat or dog.

Knowing this, I decided to exaggerate the ears to make them large features that would be used for expressive feedback. This feature also corresponded to our goal of having curiosity as a cornerstone of the robot's character, boldly announcing its intention to listen to commands in order to learn how to accomplish new tasks, so before any interaction even takes place, the physical form serves to establish its role as a social actor in the room.

Prominently placed on the upper hemisphere of the head, the ears appear as two long half capsules, almost like thick antennae. Since part of the goal of the design was to set the expectation of the robot as a machine (as opposed to a humanlike creature), the ears are more like extensions of a helmet than organic elements. They each have two degrees of freedom, allowing them to both pivot up and down and spin to face forward or backward. The result is a wide range of motion that can be harnessed to enhance the interaction. While working with the team, I learned some of the ways that a seemingly frivolous or purely decorative form serves to enhance the interaction.

The ear position can indicate many aspects of the state of the system—for example, sleep, wake, standby, busy, or in an error state. It can show the robot's successful localization of the sound of an individual's voice, something we humans do intuitively but that requires programming and processing for a robot to do well. It can show attention—that is, point toward the person with whom it is actively engaged, even if the voices of others can be discerned nearby.

Some of the most nuanced aspects of the interaction with the robot are its ability to express communicative gestures, and the ears play a big role in emphasizing the messages that the robot communicates. For example, during an exercise in which the robot was trained to identify the colors of objects through spoken prompts, the ears can pivot up and forward to indicate that it is in the process of listening for words in a sentence that relate to colors. If someone says, "It goes in the green bin," Simon can parse the sentence and pull out the word *green* and then learn to associate that word with the color of the image its camera picks up for that object. If it isn't able to discern color information from the way the person phrased the sentence, it can lower its ears, as if to say, "I didn't understand you and am trying to listen for more information." If it places the object in the wrong bin, the person can chastise Simon, and the robot can lower its ears as if to say, "I'm sorry I messed up. I care about getting this right."

Because of the robot's ability to gesture, the exchanges can be loaded with a great deal of information that is carried by the form and the gesture and therefore doesn't need to be explicit. In other words, Simon can say, "I'm sorry" in body language without having to add more words and sentences to the conversation. And while all this is taking place, the communication can serve to establish the robot's personality, indicating that it cares about doing the job well and is invested in the task at hand. Having a personality that can be understood also helps people to build reasonable expectations for how the robot will behave in future interactions. This, in turn, provides an intriguing emotional hook that encourages people to continue interaction and have empathy for the robot as an imperfect entity that needs to mess up in order to learn.

This powerful sensation of having an emotional exchange with a product is the holy grail of product design; an instant, intuitive interaction can make the difference between a person feeling frustrated with their product and adoring it. The key to pulling off this type of interaction smoothly, in which the person can be drawn into the exchange with the product and linger for as long as is needed, is not a humanoid form. Instead, the core part of the experience relies on designing a system that can respond appropriately to the exchanges that are likely to take place. While sensor systems and programmed actuators are intrinsic parts of what's needed, it all starts with the product's architecture and a physical design that's sensitive to the social potential of the interaction. So in Simon's case, it may be exaggerated antenna ears, whereas for other products it may be another characteristic, like a wagging car door handle or a microphone that can bow and point to where it's listening.

While Simon is what I call an "extreme" interface, exhibiting many sophisticated behaviors and features that are not necessary for most household products, I have drawn on my experience with it, as well as subsequent robots developed with the Socially Intelligent Machines Lab (which later moved to the University of Texas at Austin) and Diligent Robotics, a company that was an offshoot of one of the lab's projects and for which I served as head of design. I rely heavily on these insights when embarking on the design of all kinds of products, from interactive water bottles to car interiors.

Interaction Intelligence

From a technology standpoint, products and services have been augmented with the ability to speak and listen using normal language, the capacity to remember what you did last week, the potential to sense factors like the weather, and the connectivity to get updates from the internet. These feel like advances in the intelligence of our devices because we are going beyond buttons, knobs, and screens, and we have just begun to interact using intuitive human behaviors such as conversation and gestures. Interactive agents such as Apple's Siri, Amazon's Alexa, Google Home, or Microsoft Cortana are artificial intelligence (AI) agents, though the experience they enable goes well beyond what AI alone can offer. From a consumer standpoint, *interaction is intelligence.*

The distinction between interaction intelligence and AI is an important one to understand. Today's AI enables rapid search, pattern recognition, complex planning, and massive data processing. This can be helpful to making interactively intelligent systems but is far from enough to accomplish that job alone. The ideas in this book focus on the interaction intelligence that will be the cornerstone to designing socially savvy products and establishing paradigms for the next generation of consumer device design.

From a business perspective, it will be critical to hire people who can understand both the technologies of interaction and the psychological and social rules that govern people's responses to different interaction designs. These next-generation interaction designers will need a broad suite of skills beyond those taught to designers, engineers, or computer scientists today.

Modeling Interaction

In addition to the social design framework, which will be used to organize the overall structure of the book, it will be helpful to refer to a model to help illustrate basic interaction genres and analogs. This model diagrams the communications between a person and product—we'll call them the *interactants*—beginning with a common communication medium such as a phone. It is something that will be expanded upon in stories and examples throughout the book.

This book addresses how we as designers need to approach the design of interactive objects, knowing and understanding that people may interact with those objects like they are people.

A Nested Approach

While it's simple to understand the need for social design to be imbued into the product creation process in its early stages, actually implementing interaction intelligence is a complex undertaking, given the many facets of the design process and

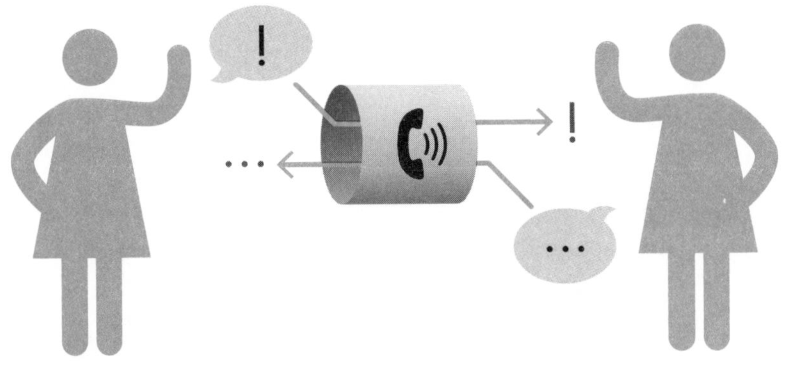

FIGURE 2-2

Product as Person-to-Person Conduit

Here, we diagram the real-time conversation two people are having over the phone. The phone functions as a social medium, a channel through which social interactions occur. The message is a communication between the two people, and it is transmitted verbatim so that what one person says is precisely replicated in what the other person can hear.

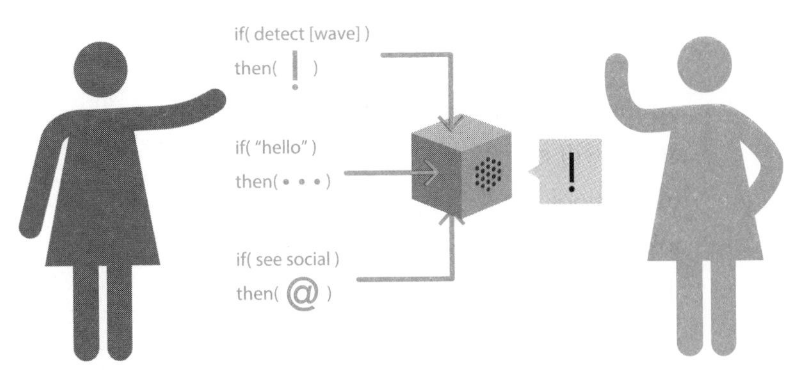

FIGURE 2-3

Product as Communicator of Embedded Messages

When a designer creates an interactive object, such as a microwave oven, certain messages need to be programmed into the object to be delivered at an appropriate time based on that person's interaction. It might be something like "Your timer is up" or "The meal is ready." The microwave conveys social messages and responses to the end user, but instead of the message being sent from a person, it's essentially sent from the designer *through* the microwave in response to an anticipated sequence of events. It is perceived by the person as machine generated.

FIGURE 2-4

Product as Independent Social Entity

More commonly, people perceive of the product as a social entity unto itself, communicating with the end user directly. If they are happy or offended when they use the product, they are happy with the product or offended by the product, not the company or the designer, of which they have little awareness. This is becoming more common with the development of highly responsive, connected products and the rise of conversational agents, but we would argue that at some level this has always been the way people have engaged with products.

the various disciplines involved. Much more complex than drawing up plans in illustration form or even three-dimensional renderings, social design requires envisioning a product from its shell through to its "soul," so to speak, or the way that it will behave, react, and relate to a person. Keeping these many elements in mind at once is the social designer's challenge, and a helpful way to conceive of a product in this holistic way is to map out the product goals, beginning with its physical characteristics (presence); considering its dynamic behaviors (expression); planning out its dialogue with a person (interaction); maintaining a sensitivity to location, timing, and state of mind (context); and positioning it within a larger network of related products and services (ecosystem). This formidable task involves sketching, storyboarding, role-playing, physical prototyping, technical experimentation, and many other concurrent activities. Keeping a team aligned and focused on the core idea of social design is complex but essential.

The socially intelligent product framework serves to assist in envisioning these many aspects holistically, with a nested structure that starts at the most physical aspects and radiating layers that build upon one another in increasing complexity. The first nugget of product vision, triggered by a product's presence, or what roboticists like to call *embodiment*, is where the next chapter will kick off as we explore how the framework relates to core design activities.

3

Product Presence

Form Follows Feeling

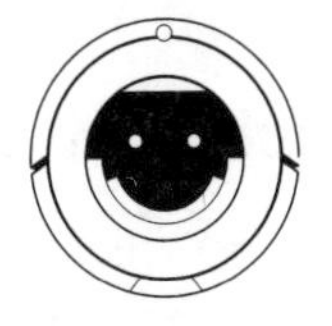

As much as I love my mother, it would be dishonest to describe her telephone habits throughout most of my adult life as anything but a pain in the neck. In my twenties I came home from work to listen to multiple answering machine messages: "Carla! I called you an hour ago! Why didn't you call me back?!" In the 1990s, when cell phones became more commonplace, I got multiple calls a day from my bossy Italian mama, and as a dutiful daughter I put up with it. About eight years ago, I introduced her to the concept of emails and the internet through the gift of an iMac. As an intensely intellectual person who could spend hours poring over the *Encyclopedia Britannica*, she was totally into it and quickly got the hang of browsers, web surfing, and *New York Times* video features.

A few years later, as the glimpses of dementia began to reveal themselves, my mother slowly lost her ability to use the computer. Email became a challenge, and then the burden of simply clicking on a browser link became insurmountable. Now, at ninety-two, the complexity of a cell phone is too great and, heartbreakingly, dialing a number on any kind of phone too challenging. In reaching this difficult moment with my mom, I could suddenly see what I'd taken for granted in terms of cognitive hurdles and how her need for something tangible and present in the room was essential.

Staying in Touch by Staying in Touch

When the Covid-19 pandemic hit, visits to my mom at the assisted-living apartment were prohibited, so her facility purchased iPads. Upon request, Zoe, my mother's caseworker, will take a break from everything else she is doing to manage paperwork, logistics, and health-care strategies and suit up like a lab worker from *Breaking Bad* to enter her room and hold up the tablet so she can videoconference with me. But since Zoe's time is scarce, it means that the video calls are difficult to book and can't happen very often. Given all the circumstances in my mother's life right now, a robotic product that would facilitate the video calls makes a lot of sense, which is what made ElliQ so intriguing.

ElliQ is a physical product that consists of what would best be described as a robotic head perched on a platform beside a tablet computer. Imagine a rotating/pivoting lampshade on a side table that can talk, nod, play sounds, spin around, and light up. Even to a robotics aficionado like myself, that description has an air of the absurd. I can hear my inner voice now saying, "Hey, wait!

I need a disembodied robot head on a table to talk to my mother? What has this world come to? That sounds elaborate, expensive, and ridiculous." However, if I put my bias aside and look purely at the core needs of the situation, it makes a ton of sense. I see in my mom an isolated older person acutely in need of interaction with family members but unable to overcome the cognitive burden of successfully navigating a tablet interface. The ElliQ is described by its manufacturer as "a friendly, intelligent, inquisitive presence in older adults' daily lives—there for them, in their corner, offering tips and advice, responding to questions, surprising them with suggestions—a dedicated sidekick on their journey through this remarkable part of life." When a call comes in, the robot head will light up and announce, "You have a video call from your friend Esther. Are you ready to video chat with her?" It will pivot toward the tablet to bring attention to it as an object and will await an answer. If ignored, it will prompt the person with reminders such as, "Esther is still trying to reach you. Do you want to sit in front of the tablet and talk now?" It can understand the person's response, and the software engine behind it will continually take into account a number of factors to understand more about the person's overall health. A number of refused calls by a person who is otherwise very social, for example, might indicate a situation that warrants examination.[1]

What ElliQ offers that so many apps and software products will never achieve is what interaction designers refer to as *embodiment*. This literally means giving a physical presence (a "body," if you will) to a software agent. Embodiment provides value for many reasons, such as:

- Providing key physical features that relate to tasks, such as a receiver for a telephone

- Relating to the human body, such as a thermometer that needs to be placed on the forehead
- Marking a symbolic value based on its location in a room, such as a bowl for keys and mail in an entryway
- Having proximity to other objects, such as a lectern facing a number of audience chairs

What *Presence* Means

Considering the salient aspects of a product's physical presence—that is, the impact of its shape, color, position in the room, and tangible elements—is a fundamental design element and rests at the core of our framework for developing product context. It is the characteristic upon which everything else is built.

There is an entire field of study established to understand the meaning and application of presence in the design of interactive objects, and a conference called Tangible, Embedded and Embodied Interaction (TEI) is a biannual event for specialists in this area to gather and share knowledge.[2]

For a clearer understanding of the reasoning behind a company like Intuition Robotics investing in the development of embodiment (plastic parts, motors, lights, and other electronic components) in their ElliQ product, let's look back at our interaction model and consider my mother's relationship to the telephone or computer. For years my mom was okay with the traditional model and the telephone as a medium, but now she does need that extra nudge, so a product like ElliQ that can serve as the interface *to* which she communicates, rather than having the extra cognitive layer of being an interface *through* which she commu-

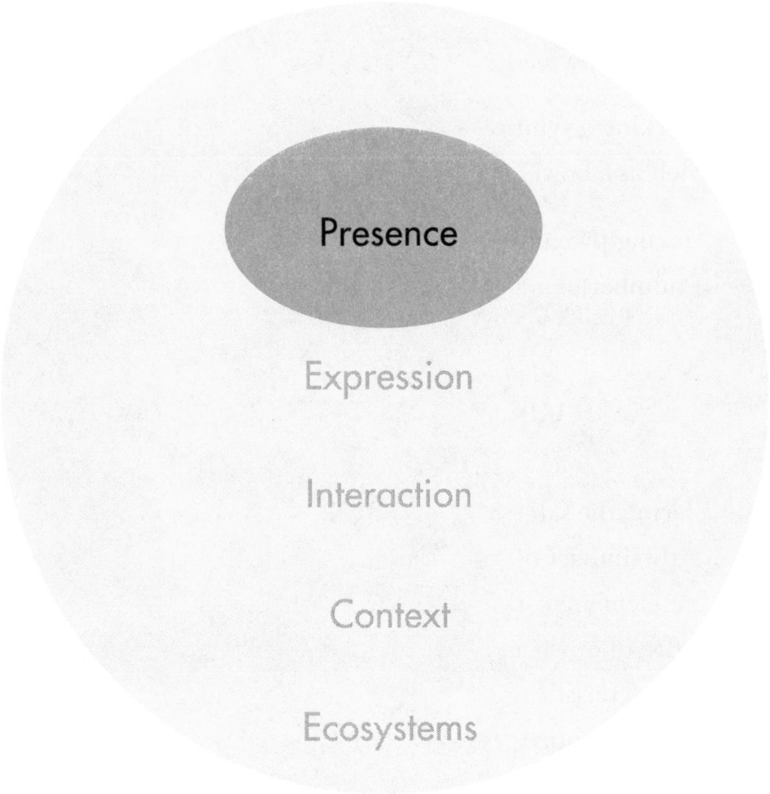

FIGURE 3-1
Presence, the First Ring in the Social Life of Products Framework

nicates, could make all the difference between her being in touch with family or being completely isolated from them and, in our increasingly virtual society, from the world at large. And the other benefits that ElliQ provides, such as medication reminders, scheduling help, messaging, and healthy living "nudges" like encouragement to stretch and go for walks, are added bonuses, but all of it is facilitated by the physical presence of ElliQ providing as streamlined an interaction as possible, allowing my mom to rest on what she knows about social interaction—someone

FIGURE 3-2

Intuition Robotics' ElliQ Personal Assistant

(in this case, something) speaks to me, I speak back—rather than having to input numbers, touch buttons, or follow prompts before she can get through to a call.

This is how we interact with the telephone. There is a great cognitive load involved in making or receiving a call that we simply take for granted but that can be a big hurdle to someone with cognitive impairments.

By allowing the person using the product to interact directly with the device, we eliminate the cognitive load involved in having to translate the need to make and receive calls into product interactions. Instead, the person can fall back on familiar behaviors of social interaction.

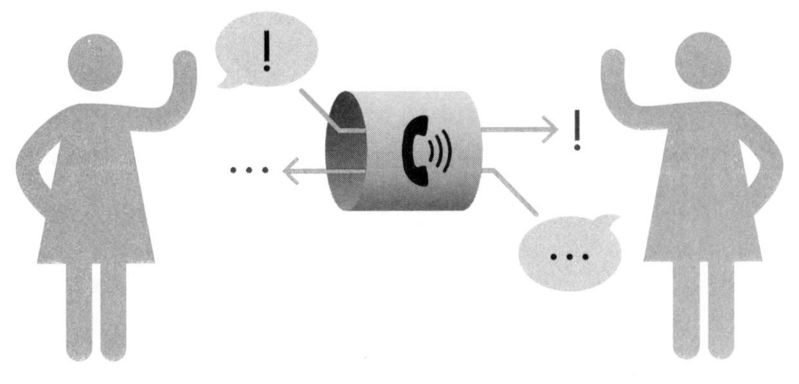

FIGURE 3-3
Product as Person-to-Person Conduit

FIGURE 3-4
Product as Independent Social Entity

As of this writing, ElliQ is still in product trials, so I haven't had a chance to try it out with my mom, but I have changed my tune from wondering if this product is a silly indulgence to seeing the light of a last chance for her to maintain communication with loved ones, particularly during the pandemic.

Product Semantics: Showing Up and Playing the Part

"The most important part of X is showing up." This is a mantra that is widely used, with *X* being a variety of things such as parenting, succeeding in school, building your professional network, and more. While trite, it is largely true for many challenges in life. The same is also true for many products, in that their presence plays a big role in their performance as social actors. This refers to their entire holistic presence—or what artists like to call the *gestalt*—as well as the details within an object.

To consider the nuances of presence in our interactions with social products, let's return to our doorman scenario.[3] In addition to physically opening the door so that guests don't have to exert the effort, his presence serves several functions. His gestures welcome visitors so they are assured they have arrived at the correct place and are given the message that they are invited and expected. For residents, interacting with the doorman on a daily basis provides a reminder of the level of luxury of that particular residence, thus bolstering the brand of the building. For would-be intruders, his presence serves as a deterrent. Not all doormen are burly bouncer types, and although they might be easily physically overtaken, their stance tells people they mean business, and at the very least, someone looking to break in is given the message that they would have a struggle on their hands. Finally, the doorman's absence at the door exterior can send the message that the door is closed because it is after hours.

The actual physical forms that are manifest in a product can communicate in powerful ways, even if their communication is implied and not consciously read or translated. Similar to the doorman, an airport kiosk can be the first interaction a person has with an airline on the day of travel. It allows people to perform flight check-in procedures, but before a person even gets that far, its presence provides travelers with the assurance they'll be able to access their trip information and leapfrog the lines of customers waiting for assistance at the desk. Kiosks are typically designed to have a strong presence; they can be recognized at a distance; they stand upright at shoulder height; their screens face upward, allowing for a comfortable view of what's on the screen; they are at an adequate distance from one another to allow privacy during interaction. Each of these aspects is carefully designed with a sensitivity to how the object's presence will affect a person's exchange with it. It serves as a representative of the airline and thus must portray the correct brand values—even the overall shape, whether elegant, tall, and thin or cute, stocky, and rotund, will communicate brand values—and it also serves as a comforting presence within the context of the overwhelming airport environment, a space that can often feel alienating and isolating.

When creating a product, a designer selects shapes, colors, and materials to deliberately communicate important aspects of the product's value and establish its character. These design details may include clues relating to how one might operate it, such as buttons, knobs, or dials, where it should be placed in an environment, and what memories or emotions it is intended to evoke.

This nonverbal language of design is often called *product semantics* and refers to the creation of deliberate formal characteristics that communicate one or more aspects of a product's role.

Design theorist Klaus Krippendorff defined product semantics as "the study of the symbolic qualities of man-made forms in the context of their use and the application of this knowledge to industrial design. It takes into account not only the physical and physiological functions, but the psychological, social and cultural context, which we call the symbolic environment." And he explained that "the designer could be cast in the role of a communicator whose messages to the user concern the symbolic qualities of products. Just as a journalist creates informative messages from a vocabulary of terms, so could a designer be thought of as having a repertoire of forms at his disposal with which he creates arrangements that can be understood as a whole in their essential parts and that are usable by a receiver because of this communicated understanding."[4] We can take this one step further and describe the designer as editor—that is, someone capable of understanding the potential of communication of a variety of diverse forms and selecting the one most appropriate to the situation at hand.

When bringing social design to bear on product creation, this means using formal and dynamic characteristics that indicate how an object can offer important cues about how people can interact with it. Aspects of its form can indicate that it can be spoken to or that it can see text, graphic codes (such as bar or QR patterns), or gestures. Designing forms that clearly set expectations for how and when to interact will enable the smoothest social interaction possible.

"The Best Camera Is the One That's in Your Pocket"

One thing I enjoyed most about my time at Smart Design was the company's culture of creating successful designs by bucking trends. One example of this was the Flip Mino HD camera from back in the days before smartphones, a pocket-sized video camera with an iconic big red button. When building the design strategy for this product, my colleagues thought deeply about the whole experience of shooting videos, from sporting events to family parties to baby's first steps. While every other manufacturer was busy pouring resources and time into developing video cameras that competed in the one-upmanship of feature specifications (higher resolutions, variable file formats, special lens configurations, etc.), they went in the opposite direction and presented people with *fewer* specs. Their research showed that the current products were relatively heavy cameras that required a series of decisions to use; people had to choose resolution, light setting, file format, mode, and more. So they challenged all of it by creating an ultralight, slim, bar-shaped camcorder that needed just one control to use, and that control would appear in the form of a big red button. To use it, you simply turned it on and pressed the button to stop and start recording. That's it. Downloading videos to a computer was a similarly streamlined process. Instead of having to track down a connector cable, you could extend a built-in USB plug and use software that was built into the device.

What might have seemed like a kids' toy was a runaway success that created an entirely new category of product. Other manufacturers rushed to copy the product's physical style and overall architecture, but the Flip remained the winner of its category for many years. The product's physical presence offered a streamlined experience that met the needs of its core customers, and because it was durable, it was more likely to be handy in a pocket when there was something fun to shoot and share. Its rapid-fire red button operation meant the video was more likely to be captured spontaneously. In terms of social affordances, this served as a prosthetic.

Although Pure Digital was bought by Cisco and the product terminated when smartphone video capabilities caught up to what the Flip could do, the success of the design was a lesson that emphasized how profoundly key decisions about the fundamental architecture of a product can affect the overall interaction.

Product Character and Product Story

While so much of the excitement around the newest genre of smart products revolves around how they behave, like the chime of a washing machine or the display on a microwave oven, their social essence begins with their visceral, physical form, driven by an overarching strategy that guides the product's creation. Designers commonly refer to that strategy as the product' *story*. It is manifest in powerful ways through form and materials and enhanced with sounds, lights, audio and text messages, and movement, even if it is implied and not consciously read or translated.

Simon the robot's story, for example, is based on it being a helpful and trainable server/sidekick that is humanish but also clearly robotic. As it was meant to focus on learning, it's got a "toddler aesthetic" to communicate the idea that it needs to be taught even the simplest of ideas, such as "What is a cup?," and "What color is red?" This manifests through a large head and wide eyes but a hard plastic robot-like shell and geometric forms.[5] Story is intrinsically linked to presence and can become a conceptual litmus test of sorts to guide design decisions. For example, we wondered if Simon should have hair but ultimately settled on a helmet-like structure that had the semantics of a robot head but could also be the sort of thing a kid would wear.

Great teams work together on developing story so that design decisions can be aligned from the start of the process through product development, building a sense of character. Defining the character of a product is at the heart of the story for social products and can be extrapolated upon to make design decisions about the details of dynamic characteristics later on in the design process.

Getting into the Heart of Things: Social Roles

Reflecting on the value of presence in the objects around you will reveal the significance of how specific forms communicate, but at a deeper level, the true emotional value comes from looking at the social roles that objects play in our lives. Sure, we enjoy the tape measure that helps us remodel the closet because its markings are clear, and it snaps back into a compact package when we're done, but we might also love how it feels to walk down the block with that sturdy tape measure hanging from a pocket: its shiny chrome and serious graphics let everyone know that it's the one the pros use; we might also love it because it was a hand-me-down from a favorite cousin who was a cabinetmaker. When creating products, we can't necessarily dictate all of the complex emotions that will go into building a strong connection between

OBJECT LESSON
Nest Outdoor Camera

Sometimes the character of a product comes easily, as a natural by-product of its task. An ambulance, for example, needs to demand attention and express the serious nature of its trajectory as it travels down a busy street. Its character is strident, serious, and authoritative. A security robot in an office setting, on the other hand, may need to interact with a large number of people in a very casual way, limiting interaction that would call attention to itself so as to avoid contributing to a hostile environment in the office. Its character might be conscientious, methodical, and self-effacing. During the research for this book, my friend and former colleague Rocky Jacob shared an anecdote about his passion for controlling the character of a design through its physical attributes, even when that means making more work for the engineering team to match the vision.

As head of design at Nest, Rocky and his team were tasked to reimagine what a new outdoor camera could look and feel like. The desire to make security feel more friendly and nonthreatening quickly became the leading paradigm. The Nest brand was based on providing customers with calm reassurance that devices were doing the hard work of keeping homes safe and running smoothly, and the camera would join the family of company products, along with the lovely glowing thermostat dome, the even-keeled smoke detector, and the sculpted indoor camera. Because an outdoor camera can benefit from a shield to minimize the impact of sunlight, Rocky started with engineering specifications for a product that would feature a cover above it, making it match the archetype of a stern and scary security

FIGURE 3-5
The Nest Outdoor Camera

camera. He felt strongly that the shield would create an overly serious character, making people think of law enforcement and the fear of crime, whereas the company wanted the product's story to be much more about proactive remote vision of the home. Think friendly uses, like checking on packages left at the doorstep and pets frolicking, rather than burglars and "Keep Out" signs.[a]

"People identify a security camera is by its 'police hat,'" he explained, referring to the sunshade that is typically built into the form. "And obviously there's a functional purpose—it's about protecting the lens from sunlight. But investing in changing the perception of traditional security cameras became a mantra the industrial design team had strong conviction around; we felt that this was the visual element that could really make it *look less* like a security camera, and we wanted to make it a little bit more inviting and approachable, because most of the time this camera's just going to live there . . . as part of the architecture. . . . But rather than being a beacon of fear, [we could] transform it into an object that provides you with super power computer vision, enabling features like tracking the health of your garden or seeing what your pets are up to." Getting rid of the visor may seem like a small move, but it required numerous technical and engineering challenges to design a camera lens that works in the various lighting conditions that outdoor cameras encounter. Nevertheless, it was critical to maintaining control over the aesthetic and character of the product. In the end, the entire team at Nest agreed that eliminating the "police hat" led to the creation of a product that felt much less threatening and fit much better with the brand than the early iterations of the design.

Beyond character, there are many functional needs that drive physical design decisions. Areas on a product that can be touched or grabbed can similarly be indicated by the presence of details such as handles or textural changes like ribs or bumps. And while it may seem extreme to craft three-dimensional shapes to serve as indicators for interaction, these forms not only are more satisfying because of the visceral nature of their presence but can also help people who may be visually impaired understand how to use objects and navigate their environments. The sidebar on "Mental Models, Mapping, and Affordances" describes some tools derived from cognitive psychology that designers can use when envisioning product details.

a. Rocky Jacob, interview by Carla Diana and Wendy Ju, audio recording, New York, NY, March 6, 2018.

a person and product, but the best designs stem from gathering as much understanding as possible about the potential for connection.

In this section we describe some key factors in the social nature of objects, beginning with three aspects of the overall relationship we have with our products. These can be generalized as:

1. *Relational:* What role does the object play in the relationship between person and product?
2. *Emotional:* What feelings does the object evoke through its use?
3. *Conditional:* How do the various states of the object affect the way it's perceived and used?

Relational

Whether we are aware of it or not, products play social roles in our lives throughout the day, though some are more pronounced than others. Acknowledging the role can serve as a metaphor upon which to build design stories, guiding details in expression and interaction (which we'll get into further in subsequent chapters) based on behaviors that people already know and understand. Products may relate to people by taking on the role of prosthetic, tool, assistant, vehicle, or even placemaking. As designers, we consider how the social affordances define the degree of intimacy one has with a product: Is it more like a prosthetic that is used as a tool or extension of the body, or more like an agent that acts on one's behalf?

When designing a floor-cleaning robot at Smart Design, my team and I recognized the relational value that a floor-cleaning robot might have. In addition to it entering as a new product archetype into the intimate environment of the home, it was introduced at a time when it was likely to be customers' first experience with an autonomously navigating product. In thinking about its social role, we set out to create a device that would be akin to an assistant. We dissected typical service personalities as a starting point for what the product story would be:

- *The hotel housekeeper:* Quiet, conscientious, and for the most part a person with whom you don't interact at all. The best housekeeper is one in which you see visible evidence of the work she's done but you don't actually see her.
- *The butler:* Proper, obsequious, and attentive. He won't let on to what he's really feeling and will for the most part hide all emotional responses but makes you feel comfortable requesting almost anything.
- *The bartender:* Chatty, curious, and empathetic. He feels like your best friend even if you've only known him for five minutes.

- *The nanny:* Traditional, nurturing, and rule-bound. You feel safe when you're around her, but you also know she'll correct you if you go past the limits she's set.

We then used some of the attributes that emerged—precise, hardworking, intelligent, and humble yet playful—and modeled the interaction around the unflappable inventor's companion Gromit of *Wallace and Gromit.*

Emotional

Everything we encounter has some emotional significance to us, though it's more powerful for some things than others. While it may seem overly sentimental to look at products through the lens of emotional significance, objects carry enormous meaning for people for a variety of reasons. Even something as seemingly mundane as a light switch can offer power or leave someone feeling vulnerable depending on the nuances of its design. For example, it may seem like an oven's only purpose is to heat food, but every detail of the oven is an opportunity for a social relationship between the person and the product as well as the person and other people around him or her. The shape of its knobs may reference restaurant kitchens, making a person feel like an accomplished chef; the oven door might match the interior decor, blending into an environment that encourages people to use the kitchen as a gathering space; the typeface on the interface might have a retro feel from the 1950s, making someone feel nostalgic for their grandmother's pies. Designers as well as marketing professionals know intuitively how powerful these emotional cues are, and in order for everyone on a product team to maintain alignment, it's important to acknowledge, and in some cases elevate, a product's emotional role for the person who will use it.

Below are a few metaphorical categories to consider when thinking about how the product you're creating will evoke emotion for the person using it.

- *Totems* are objects that represent power and the potential to fulfill a need. The airport kiosk mentioned above is a good example of a device that fills this role. An audio speaker is also a great example of a totem—it reminds its owner of the importance of music in his or her life, and its details, such as a traditional hardwood material versus a slick black plastic, reinforce that person's identity.
- *Talismans* are similar to totems in that they provide a connection to a power beyond the physical artifact but are different in that they are typically smaller and can be held or worn on the body. A wrist-worn activity tracker like the Misfit Shine can serve as a magical object that propels someone to better health through both its ability to count steps and connect to a tracking interface, as well as serve as a symbol of the power of taking control of one's habits.
- *Badges* are products that reinforce identity. A key fob may function to open a door but in many cases will serve a social role as well. When consulting for a leading private jet company, my team and I recognized that many plane owners enjoyed belonging to an elite class. Having a key fob that was not only functional for the person using it but visible to others as well served to reaffirm their identity. This sense of identity with a "tribe" can apply to many categories of products. Even a product as simple as a water

bottle might serve as a tribal badge, communicating a sensitivity to environmental concerns based on the shape and material that it's made of.

- *Mementos* are objects that remind us of others and are arguably difficult to intentionally define. I still have a wristwatch that my late father was given upon his retirement decades ago. A vintage Timex from the 1970s with a personal inscription that says, "Best of luck, Joe," it bears no resemblance in style to anything else I might wear or use; however, it carries enormous emotional significance for me and therefore trumps any other kind of wristwatch that I might wear in terms of being meaningful. While we won't always be able to predict how objects will serve as mementos, it is a powerful category to keep in mind when developing a product story.

The psychologist and author Mihaly Csikszentmihalyi has written several books about human emotion and personal fulfillment. He wrote *The Meaning of Things*, a study of the significance of material possessions in contemporary urban life, which has served as seminal reading to help product designers understand the depth of the connection.[6] In fact there is an entire field of study in material culture that delves deeply into emotional significance and can serve as a strong starting point to understand the relationship between a person and a particular artifact.[7]

Conditional

Conditional character traits are the definitions of how a product behaves in multiple conditions, such as off/on, sleeping/charging/active, open/closed, connected/offline, and so on. While understanding that the emotional and relational aspects of a product are fundamental to setting the foundation for a product story, the conditional aspects are what allow the story to evolve, and considering all possible conditions can help a character feel holistic. The airport kiosk that functions as a totem, for example, should continue to do so even when it's in a sleep state through its architecture, graphics, and perhaps even a light that gently glows.

Laying out all the conditional possibilities for a product will greatly inform initial product definitions that can be used at the onset of a project. While this is a time to lay out the bare-bones functional requirements, it is also an opportunity to build in the necessary behaviors that will round out a character, taking into account all the states it will be in over the course of a person's interaction with it.

In addition to solid form, electronic objects have the ability to show changes in condition through dynamic characteristics, such as light, sound, and alternative shape configurations that emerge as a result of movement. For example, designer Naoto Fukasawa created a vacuum cleaner concept model for Hitachi with a light that would remain largely hidden until the vacuum was full and needed to be emptied, at which time it would begin to glow in its center, as if a full belly was emerging.

Let's return to the floor-cleaning vacuum design project described above. Once my team and I had settled on a character definition for the robot, we identified the extreme moments of interaction: When was the robot happiest? When was it most troubled? Just as a person's character is revealed by how they react in these extreme moments of human experience, so, too, will a robot's character be perceived. To actually apply this character definition, we created a language of expression that combined light patterns, a sound palette, and choreographed movements. For example, getting stuck under the couch would be a moment of distress, as would having its batteries running low. Completely cleaning a room's carpet might be a moment of jubilation. We wanted to define exactly what the robot would do at these moments. What sounds should it make? How would it move? Just as we might work with a color or material specialist on a traditional product, here we enlisted a music composer to create a palette of sounds, as well as a visual designer to create a custom font for the LED matrix display. To match the "Gromit" character we had chosen, we built in just a little bit of goofy playfulness, like a visual icon with the word *Yum* when it passed a particularly dirty spot, or a greeting tone if it recognized a person in its path, as if to say, "Oh, hello!" Moments of distress, like being stuck under a piece of furniture, were handled, in the manner of Gromit, with grace, with tones that expressed trouble but not extreme anxiety—something more like "Uh oh!" than "OH NO!" Its movement was planned to match these characteristics. In the moment of greeting a person, it does a little wiggle backward before continuing forward to do its work.

When looking at how motors might move parts to modify a product's form, like Simon's ears or ElliQ's pivoting head, we can consider how shifting stance can create meaningful and salient conditional movements during interaction. Pixar animator and robotics consultant Doug Dooley describes the importance of stance in terms of how certain parts of the robot's anatomy are positioned with respect to one another. He explains, "If you want a person to feel as though the robot is interested in what they are saying, the robot needs to lean in toward them for an engaged stance. If a robot is to appear embarrassed, the robot should probably lean away in a disengaged stance. . . . You will sometimes hear animators discuss this concept as an 'open' or 'closed' pose." He also said, "I show a character's confidence by arching the character's chest out, or arching the character's chest concave in. Since the pivots of the spine are in the back, this also controls how tall and or slouched the character is."[8]

Even an object that is not electronic will alter its presence based on how we interact with its physical components. The 1969 Olivetti Valentine typewriter, designed by Ettore Sottsass and Perry King, presented a radical departure from any past typewriter design in its slim profile and daring red plastic shell, but perhaps the most important aspect of its presence was the travel cover that housed the machine. When removed, it lent itself to being placed on the floor so that it immediately transformed into a trash bin. The presence of this accessory encouraged the passionate disposal of crumpled up and rejected drafts of novels, poems, and love letters.[9]

Social Cues and Why Clear Communication through Physical Architecture Is More Important Than Ever

Presence is particularly important to communicating the social roles of a product since social interaction is often about potential. When it comes to specifically human interaction (upon which we base our understanding of social exchanges), we gauge what's socially possible and appropriate by presence. At a party, for example, a person's physical presence will afford or prevent an introduction. You wouldn't yell at a person across the room; instead, you might approach that person slowly and physically make your presence known by standing in front of him or her. Your position

OBJECT LESSON

Clikbrik Drummer's Metronome

In 2015 my former Smart Design colleague Ted Booth described a product that his neighbor wished existed. "Konrad is a professional drummer and can't find a metronome that works for him." He explained that the drummer sets the tempo for the whole band, sometimes even guiding people through eye contact and head nodding to make sure everyone is feeling the same beat. "Every single metronome on the market is a fussy little electronic device that requires him to hunch over a box and press tiny buttons. I wonder if we can design something that lets him stay in the moment and keep his presence as the rock star performer that he is?" We put together a team and set to work, and the resulting product was the Clikbrik, a metronome that is operated entirely by drumstick; you strike it once to start the rhythm and again to pause it. The dial for changing the tempo has notches designed to hold the drumstick tip; it has a fitting that lets it screw onto a stand to be part of the drum kit, and the display is made up of large LED lights so it can be seen from wherever the drummer is sitting. Our small team was able to prototype, develop, patent, and produce the product, and Konrad and drummers like him love how the device lets them maintain their drummer's stance and persona, maintaining the stage presence that is so important for connecting both with the audience and the rest of the band.[a]

While these humanlike social cues—in the case of the Clikbrik, taking a small box and buttons and changing their size, shape, and position—may seem like frivolous additions to a product that will increase the cost without functional benefit, let's take a step back and think about the effect that miniaturization and immateriality has in the products that we commonly call smart, such as smartphones, smart speakers, and smart doorbells. In addition to the complexities of simple interactions such as turning the device on and checking its status, there are dozens of other interactions that people may not even consider. While people may know that cameras and microphones are being used for the benefit of smoother product interaction, the fact that these elements are hidden within forms that belie their existence is doing a disservice to both the person using it and the manufacturer.

The first-generation Amazon Echo, for example, sits silently and inconspicuously, only lighting up when it has been summoned, yet as people have become more aware of privacy concerns with devices that have embedded cameras and microphones, they have developed a growing list of questions and concerns about what's really happening with the device from a social point of view. The Amazon Echo embodies the Alexa assistant in physical presence; shaped like a small cylinder, it is largely designed to disappear into a corner except when summoned with its trigger word. When summoned, it does a good job of letting people know that it's actively listening, with a moving highlight on a glowing light ring it points in the direction of the person it's listening to. When idle, however, it does a poor job of letting people know what it's doing from a social point of view.

If you walked into a room with a friend and there was someone already in the room, sitting in a corner, eyes cast downward, and not interacting with you, it would feel suspect. You would be careful about what you said and wonder if the lurker was listening and question his or her intentions. If the person was a friend or colleague, perhaps lost in a book or mobile device, you would still acknowledge one another and perhaps ask the person for privacy if you needed it. These same concerns are true for our smart devices, which can benefit from a great deal more social affordances than commonly exist today. The mere ability to "bow a head," for example, perhaps through forms that can pivot against each other, could give a product like the Amazon Echo the ability to use physical presence to communicate whether or not it's listening.

Many of the issues around privacy can be tackled head-on through form, and specific aspects of presence in a physical object can be harnessed to communicate several important interaction elements. The overall form of a product will communicate orientation—that is, which surface serves as top, bottom, front, back, and so on. This can be critical for indicating what direction a mobile robot might move or how someone should approach a smart device to interact. Capabilities for interaction such as hearing and sight can be communicated by the presence of physical details that embody the semantics of these capabilities, such as microphone hole patterns or camera lens rings. This not only signals that they exist but tells people where to look when a camera interaction is needed or how to direct speech if voice is an important part of the communication. For privacy concerns, offering an opaque shade that can be slid over a camera can give people peace of mind, even if it's not technically necessary for the camera's on-off control. And it may not be possible to provide proof that an embedded microphone is not on, but giving a clear formal indication of a microphone's status can be reassuring: on, off, actively listening, or on standby.

As we enter an age in which critical and fundamental human rights are being violated through the use and abuse of our data, it will be more critical than ever for designers to find ways to communicate a product's state and intent through deliberate design elements.

a. Edwin Booth, Carla Diana, Michael Glaser, Konrad Meissner, assigned to Clikbrik, LLC, "Contact Responsive Metronome," Patent 15/772517, April 30, 2018.

matters. You wouldn't stand behind someone but instead position yourself so there is the potential to make eye contact. You may even stretch out a hand as a further indication of presenting the potential to socially interact. When we feel uncomfortable in a social situation, we modify our presence to indicate our willingness to engage. We may hang our heads, turn our back, or direct attention toward social interactions through or with a handheld device such as a smartphone.

SOCIAL DESIGN TAKEAWAYS

The stories above show how people read social cues and respond socially even when a product isn't overtly interacting with them. Its mere physical architecture sets the tone and establishes the nature of the relationship between people and product. Here are a few key ideas to keep in mind at the early stages of the product design process.

- ✓ There are social aspects to every part of the design, even the physical form.
- ✓ Product architectures not only enable specific types of interaction but act as physical instruction guides of sorts, offering a mental model that guides every aspect of a product's use.
- ✓ The relationship a person has with a product can be described through a story that can guide the overall creative direction of a design project, determining the product's character and key attributes.
- ✓ Product design characteristics can be framed through relational, emotional, and conditional aspects, and a strong design will take all three into consideration.
- ✓ Just because a product characteristic can be hidden and embedded doesn't mean it should be. Today's smart products are desperately in need of more physical presence to communicate the important, yet intangible, aspects of their operation, such as using microphone and camera data as input.

Mental Models, Mapping, and Affordances

When looking at an object, a person will try to understand its use, even if they have never seen it before, and the myriad elements of its presence are at the heart of how well it is understood. A gold standard for good design is a product that can communicate its use without relying on a person to review a set of instructions. This is an enormous challenge—just looking at videos of kids attempting to use an old rotary phone can illustrate how hard it is to communicate interactions, especially those that involve more than one operation to complete a task.

Generalizing broadly, the task of making a product do its job (e.g., heating filaments to toast bread) is allocated to product engineers, but the metatask of helping people to understand how to make the product do that job (e.g., making it obvious where the bread goes in) is allocated to product designers. Designers need to anticipate what people know or don't before encountering the product and predict what sense people will make of the product as they get to know it better. Designers need to shape how people think products function by planning for the "mental model" the users might have of the product.

Mental models form a strong foundation in the very early stages of product development and therefore can be a great place for a team to start when forming a design strategy for interaction. Being able to articulate a mental model, whether through a verbal description or visual diagrams, is essential to creating something that people will enjoy using.

Two key aspects of mental models are *affordances* and *mapping*. In the design community, the term *affordance* refers to the elements of a product that signal how a person should interact with it. In *The Design of Everyday Things*, Don Norman explains that ". . . visible affordances provide strong clues to the operations of things. A flat plate mounted on a door affords pushing. Knobs afford turning, pushing, and pulling. Slots are for inserting things into. Balls are for throwing or bouncing. Perceived affordances help people figure out what actions are possible without the need for labels or instructions."[a] On a toaster, the lever that lowers the toast and starts the toasting function is usually designed in a way that "affords" being pressed down, and there is usually a visible cue that the lever moves vertically. These affordances help the user construct a mental model that they need to put the bread into the toaster and push down on the lever knob to operate the toaster.[b]

Mapping describes the way that controls and resulting actions are related. For example, on a toaster, pushing down on the lever causes the bread to be lowered into the toaster; when the toast is done and the toast pops up, the lever also pops up. This mapping of up-to-up and down-to-down makes intuitive sense; people don't need to understand the mechanical systems inside the toaster that make that correspondence to understand it.

There are other *semiotic* aspects to the toaster's design, which provide signs about the states or settings of the toaster. The lever, for example, stays down when the toaster is toasting; the lever's visible location is a *sign* of that state, even when it is not acting as an *affordance*. The knob setting the length of the toasting usually has other graphic indications—sometimes numbers, sometimes a gradated curve, sometimes just icons of white, tan, and dark toast—which signify the amount of browning.

a. Donald A. Norman, *The Design of Everyday Things* (New York: Basic Books, 2002).
b. Ibid.

4

Object Expression

Communicating Behavior

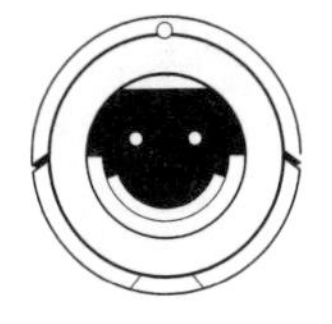

"OMG!" I can hear Pecorino, my three-year-old Chihuahua mix, exclaim. "There's someone on the other side of the door! Come here! Quick!" My dog, of course, neither speaks English words nor gives me specific directives. But he does communicate these messages very clearly through the pitch and volume of his bark, his dancing gestures, and his overall posture. I can read the pivot of his ears precisely and know where activity is happening—outside the front door or in the backyard. I can tell the difference between someone who's just passing by or someone who is lingering by observing his jumps—he'll pace around by the doorway a few times for a temporary passerby but escalate to a jumping frenzy if the person stays and knocks.

The Power of Shorthand

Pecorino and I communicate in a sort of shorthand: I can tell what he's talking about, where it is, how he feels about it, and what he wants to do next, all at once. Dogs have been bred and domesticated over millennia to work next to people. They have evolved to be particularly good at reading people's communicative cues and responding in kind.[1] It seems like mind reading, but really, it's body language.

Much like my dog, a product can express itself in many nonverbal ways, and as designers we can use this communication to quickly and intuitively offer feedback and other information. In this chapter we focus on ways that dynamic product characteristics can serve as a richly expressive yet highly efficient method of communication.

Just as we read subtle cues from our pets, we can read messages and emotion from our products, perceiving nuances of dialogue and a sense that the object is "alive" when it is actively interacting. Our washing machines can wiggle their doors to remind us to put wet clothes in the dryer. Our vacuum cleaner robot can perform a dance to show that it's proud to have finished a room. These animated behaviors blend together in a magical way, and it's human nature to decode them as if they are emanating from a living entity.[2]

As products continue trafficking in more and more complex information, the temptation to use speech or screens to communicate everything is powerful. While it's popular to think of elaborate text or voice exchanges, often called *natural language interfaces*, as the best ways for people and products to communicate, this is often not the case. These explicit modes of communication require our undivided focus and are far less useful in environments where we might have demands on our attention, such as in a car or in the kitchen. While today's Siri uses lengthy interstitial language, such as "Today's weather will be . . ." and "It appears that you are . . . ," the most powerful aspect of our relationships with our products will be the split-second, near-telepathic exchanges that can happen with just a flicker of light, a sequence of tones, or a gestural movement—the kinds of messages that can benefit from our full attention yet can also take place in our peripheral vision.

Expression

Expression forms the second ring in the social life of products framework. In the last chapter, we talked about presence, the overall impression of a product based on its embodiment, considering aspects such as its color, the shapes of its parts, and the story expressed through its overall architecture. Progressing to the next ring in the social design framework, we take an intimate look at a product's ability to use physical presence to communicate messages and respond to the people and environments around it, making apparent important details about its internal states. We'll consider ways that a product can enhance its physical form and make the most of conditional abilities. And we will expand upon ideas around the social affordances of dynamic behaviors crafted by expressing information through light, sound, and motion behaviors, exploring the wide palette for expression that's available to product designers.

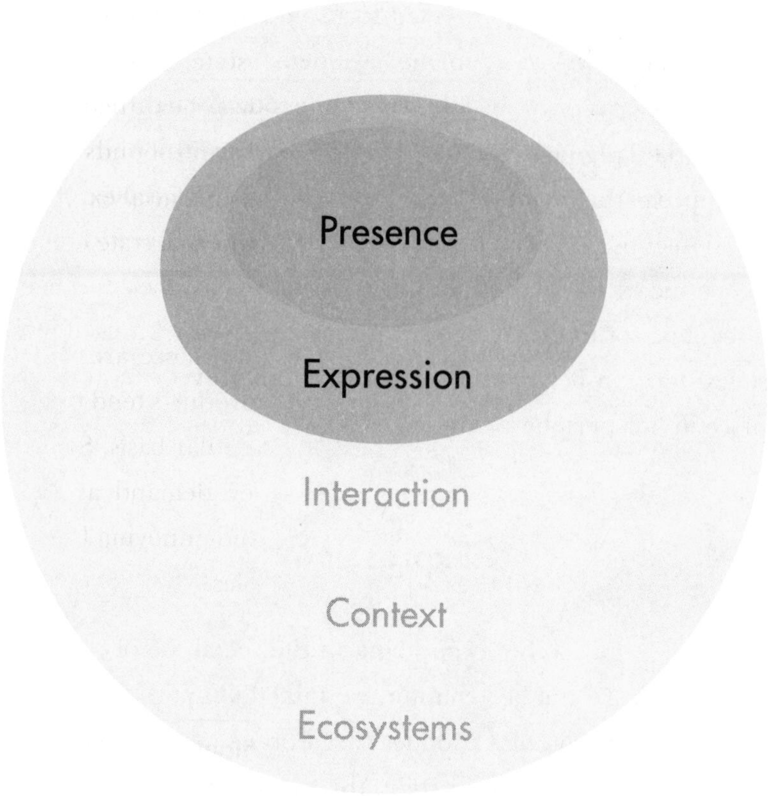

FIGURE 4-1
Expression, the Second Ring in the Social Life of Products Framework

Product "Body Language"

In *Turn Signals Are the Facial Expressions of Automobiles*, Don Norman explains:

> Facial expressions, gesture, and body position act as cues to a person's internal states. We often call these things "body language," the name indicating the communicative role. Body language makes visible another's internal state. The blush of the cheeks, the grimace,

> the frown and the smile all act as readily perceivable external signals of a person's internal state, making visible to observers what would otherwise be difficult or impossible to determine. . . . The lights and sounds of an automobile play a role analogous to the facial expressions of animals, communicating the internal state of an auto to others in its social group.[3]

Starting at the very core of a product's essence are several messages I have observed in my work that products tend to communicate to the people using them on a regular basis. Some of these things take place so often that they demand as much shorthand as possible to avoid a constant and annoying barrage of messaging. Table 4-1 shows some examples.

TABLE 4-1

Messages that products tend to communicate

"I'm alive."	Power, via cord or battery, is connected.
"I'm awake/asleep."	Standby mode status.
"I'm waiting for some more info."	Pinging a server or other data source before an operation can be complete.
"I need *you* to give me some more information."	Awaiting a person's input through interface elements.
"I heard you."	Confirming that a person has input the necessary information.
"I'm in the middle of doing something."	A process is taking place that will require some time to complete.
"I've just finished doing something."	A set of tasks is complete; product is ready for a new task.
"Something is a little bit wrong."	An error has taken place, or there is another issue affecting normal performance such as the inability to read a sensor or charge a battery in time.
"Something is seriously wrong."	An error has taken place that will interfere with performance.

With a floor-cleaning robot, these messages might be seen as a core set; however, there might be many more that take place throughout the course of interaction involving both functional needs, such as scheduling cleaning times and identifying floor areas, and emotional needs, such as celebrating the moment a house has been fully cleaned or expressing regret for needing to be rescued from underneath a couch. The tone and content of the messages delivered will contribute to a product's perceived character and therefore become important aspects of designing a product that feels holistic.

Light, Movement, and Sound: The Ways of Object Expression

As products become more interactive and content-driven, such as Amazon's Alexa or Apple's Siri, they add speech to the other modalities when necessary, but even spoken words need to be integrated into the overall expression, with the understanding that they are part of an entity that is also expressing messages through light, nonverbal sounds, and movement.

Light

When my little boy Massimo was one, I greeted him in his crib every morning to see him gesticulating with both arms toward the wall and explaining in baby talk that it was 7:00 a.m. and time to get up. While I'm tempted to tell you that my infant recognized numbers and learned to read a clock, what was really going on is that he and I shared the same vocabulary of lights: the dim blue hexagon meant it's time to go to sleep, and the bright orange-

and-yellow rainbow meant it's time to get up. Behind the scenes, I had programmed the light system, composed of Nanoleaf Aurora modular flat tiles, to match our schedule so that the colors automatically mapped to certain times of the day. The end result was a language we both understood, consisting of abstracted messages composed entirely of light. Nanoleaf CEO Gimmy Chu echoed the satisfaction I experienced, telling me that customers used the panels as "extensions" of environments to create a certain feeling, from a calm forest, to a vibrant sunrise, to a creepy dark room.

People take for granted the number of messages that we intuitively read throughout the day through changes in light. The glowing button on the coffee maker tells us when the water is heated and ready for brewing. The light on the range top warns us that it's too hot to touch. Flashing overhead lights in a theater tell us that intermission is over. Virgin Atlantic's Boeing 787 Dreamliner touts subtle shifts in ambient lighting, from rosy amber to bright blue, as a programmed way to communicate shifts in time zones both consciously and unconsciously.[4]

When driving, we use lights as an extension of ourselves to indicate when we are turning left or right. Other drivers as well as pedestrians understand this language and use our signals as the basis for decisions regarding how and where they will move across a street. Light can be used like semaphores, to convey explicit coded messages, but maybe the more powerful way that light can be used is more subtle, functioning like the blush of a person's cheek or the furrow of a brow to convey internal state or to react to external context. When mapped well it can direct attention, set context, and convey messages. It can be seen at a distance, so it is a good choice for products that may be positioned overhead or in a corner of a room to be viewed at a glance, such as situational monitors (like security cameras, thermostats, or Wi-Fi routers). It is also particularly valuable for robotic products that move around a space and may be far from the person because it can serve as a distant beacon that can communicate clearly.

The introduction of microprocessors along with the availability of LEDs has opened up the palette available to designers considerably by introducing the ability to control the intensity of the light as well as its color, enabling more complex and sophisticated messages to be communicated with a single element. Color and intensity of light add detail to underlying messages and can be mapped to specific values along a range, with red, for example, being 0 percent and green, 100 percent. The location of the light will draw attention to a part of an object, a gestural interaction, a feature, or a form detail that is critical to convey the object's use and meaning.

The combination of lighted elements also allows for sequences of intensity and color to be programmed so that they emit an expressive animation. The front indicator light on the early MacBook Pro is a compelling example of an object that used a simple animation in light intensity in order to achieve a powerful effect: the illusion of being alive. A soft glow shone through the body of the computer, pulsing gradually from light to dark at a regular rate that mimicked human breathing. The effect was mesmerizing and intuitive—so much so that Apple even patented the *breathing status LED indicator*—letting people know the computer was still "alive"—that is, the battery had power, and the system was asleep.[5]

Multiple lights can be strung together to form low-resolution screens that can be positioned anywhere on a product, following the curve of a surface or even glowing from within. When series of lights are adjacent to one another, the changes in their color

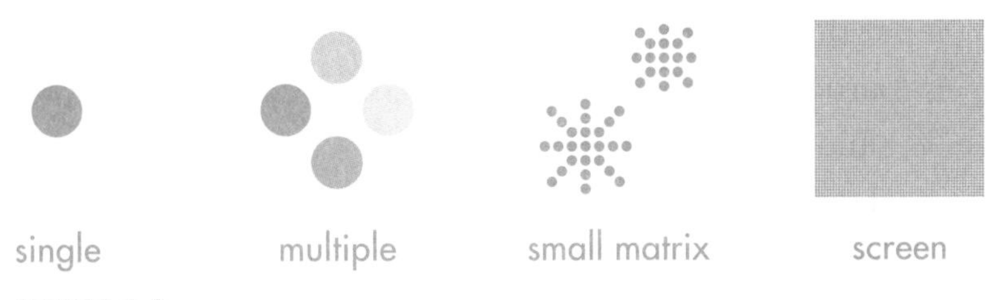

FIGURE 4-2
Arrangements of LED Lights to Place on Products

and intensity will be read as an animation that can be used to craft a variety of messages. Robotic toy products Dash and Dot from Wonder Workshop are remarkable in their use of light. Twelve LED lights positioned in a circle on the robot's face create the illusion of one eye. It can show attention by pointing the light in the direction that it's headed but can also use a sequence of light changes, from bright to dim, to indicate changes in expression. A happy response can be communicated by having all the

FIGURE 4-3
Robotic Toy Products Dash and Dot from Wonder Workshop

lights flashing around the perimeter of the circle, whereas sadness or disappointment is shown through lights that are lit in sequence toward the bottom of the circle.

Designing products involves thinking about the overall form and how light will integrate with it. Will it emphasize a certain part of the product, making that dominant? Will it create a sharp spotlight that immediately grabs attention or a gentle wash of light that's reflected onto a wall, table, or other surface? The size of the light can also emphasize certain features over others. Electroluminescent panels can glow, but they don't emit or "throw" light, so they are useful in situations where only a highlight is needed.

Light is fluid and can fill a three-dimensional space and be integrated into a surface conforming to the physical space that it fills yet still maintaining the integrity of the larger object. When we design with it, we can think about filling a form rather than "painting" light onto a "page" that can only be flat. A voluminous light can imply a space that people can enter or avoid. A spotlight may create a column of brightness and/or color to emphasize a particular area or object. The ElliQ personal assistant has a range of light patterns that glow from the robot's head, providing a spot of light or radiating rings. In a podcast interview, Dor Skuler, Intuition Robotics CEO, explained, "The light patterns are very simple cues to give the user an idea of when she's talking, when she's expecting you to talk, when she's listening, when she's thinking, when she's stuck, etc."[6] The light in this case thus serves as a dynamic material, changing the object's overall look and bringing attention to it on demand.

OBJECT LESSON

Glow Caps—"It's Time to Take Your Meds"

The Glow Caps medication bottles tackle the large problem of prescription medication compliance using an embedded light that glows to indicate when it's time to take a dose, along with an audible signal. The system also includes a wall-mounted plug-in light so that reminders appear in multiple places. This is a good example of a design that includes light features that consider multiple contexts and possible situations. When setting up the bottles, schedules can be set to map the light indicators to times of the day to correspond to when the medication needs to be taken. Since people tend to keep their medication bottles out on a counter or tabletop, having a light indicator allows it to be conspicuous. In the case when bottles may not be visible, there is also a plug-in wall light that allows the notification to take place. The lights on the wall mount and bottles draw attention to these items over others in the room. The condition of the light, whether on or off, maps to the state (medication needs to be taken/doesn't need to be taken). The lights are also designed to serve one type of user yet play a role in the larger system, which takes into account situations in which the light indication is not enough, and it will notify caregivers of a missed dose. A history of compliance can be provided to a caretaker or doctor through the back-end system.

Some things to keep in mind when using light include:

- Use light sparingly, asking the question, "Is the product's message important enough to wake someone?" When considering the number of lighted things that might be in a given room, it becomes increasingly important to avoid creating a "constellation" of points of light.

- Take advantage of all of its dimensions. Light offers a reading of value in intensity, color temperature, and color saturation.
- Animate a surface through timed responses. While a single light can simply blink, a series of lights situated next to one another can communicate many messages, such as pointing toward a direction or expressing excitement or calmness.
- Beware of competing with daylight. Light loses its impact in products that will predominantly be used outdoors.

Movement

On stage at the 2015 TEDxPeachtree conference, Dr. Andrea Thomaz gave the introduction to her talk with the robot Curi, poised nearby.[7] I had designed the shells of Curi, Simon's "cousin," to have similar architectural features, with expressive round, glowing ears and large, sympathetic eyes. As an important task for a robot is to recognize objects in order to navigate the environment and manipulate things appropriately, the demo setup for the talk showed the advantage of having an articulated robot body when interacting with a person. To show the power of combining words with robotic gestures, Andrea prompted the robot so it used its robotic voice to ask, "Is the green object seventeen inches from the table edge a flowerpot?" Next, she prompted the robot to make the same observation but use its arms and fingers to point and ask the question more succinctly: "Is this the flowerpot?" The demonstration exposed how much more natural the communication was when movement was harnessed. While the elaborate mechanism of the robot may seem like a great indulgence, if it were a device in the highly stressful environment of a hospital setting, the efficiency of a gesture such as the one described above could save valuable time and cognitive energy on the part of the person using it.

The Power of the Nonverbal. People are highly sensitive to perceived motion. Things moving in nearby space can pose threat or opportunity, so from an evolutionary point of view, our senses are finely tuned to notice movement.[8] Much of what we do to communicate verbally is presaged and augmented by nonverbal cues. We can easily catch one another's attention by waving, direct attention by pointing, and signal a willingness to engage—or not!—with a turn of the head or a shrug of the shoulders. Movement can help mediate understanding across distances, in loud environments, or between people who speak different languages.

We are also incredibly attuned to more implicit movement-related cues. Just the *potential* for action communicated through a person's stance makes for great differences in our feelings of intimacy, comfort, or threat. Formal design details such as orientation are important—we're much less wary of a stranger who is turned away from us than someone who is looking right at us from the same distance. A short bow toward a passerby, a nod to an empty seat, or a lift of the hands toward a door handle can indicate an unspoken offer to sit or to open a door. These are not canonical gestures in any sense, but we understand them and employ them without being taught. Movement is part of an intuitive language that even animals understand.

Because movement generally requires a lot more energy than sound or light to produce and because moving parts have a greater tendency toward failure, we don't often see it used for expression in consumer electronic devices. As these gadgets take on more tasks that already require motion, though, its use will become more ubiquitous. The vacuum cleaner robot, for example, already has motors that allow it to be driven across the floor; it only needs to be reprogrammed to use those same motors to signal issues or request confirmations from the people using it. "What is that, Vroomie? There's a block you can't pick up under the sofa?"

The key to taking complex communications and translating them into the elegant nonverbal shorthand of movement is to become an expert in abstraction, similar to the way that animators do. They focus on identifying key emotional moments in a film and then draw a character that can exaggerate those emotional expressions. Designers can define key moments that are likely to take place during typical product interaction and then specify the way the product should behave in each case.

Students of animation begin with a standard exercise in which they are asked to create a flour sack that's as minimal as possible in terms of form and decoration and then animate it to express a wide range of emotions, such as excitement, shame, shyness, elation, smugness, and so on. The assignment is restricted to a flour sack in order to focus the animator's efforts on the movement above all else. It's a formidable challenge, but talented animators manage to make that sack really come to life despite its lack of any real physical characteristics like a face, limbs, or head, evidence to the fact that nuanced movement can be used as a language to communicate emotional messages.

As product designers we challenge ourselves to do a similar thing using moving parts of products. If we design a simple over-

FIGURE 4-4
The Flour Sack Exercise for Design Students

all form such as a cube or a cylinder, we'll need to rely on dynamic behaviors to allow it to communicate like some of the social robots being developed in labs. Given its form, we will ultimately be relying heavily on abstracted or implied movement to convey emotional messages and may ask ourselves questions such as: How can we make a cylinder bow to express regret? Could a cube inflate itself triumphantly to show pride? Or cower to be fearful?

Asking and Answering the Right Questions. Answering these questions in a way that can aid product development requires a few stages. First, there's knowing what to do—that is, defining your messages in human terms. For example, a vacuum cleaner can let you know it's happy because it's successfully completed its cleaning or in distress because it's stuck under the couch. Next is the step of abstraction, or decoding the messages into choreographed movements, so that the "I'm proud that the living room is clean" message is a moment that leads the vacuum to move in a way that makes it appear to do a happy dance. The final step is knowing how to use using mechanical devices such as motors, actuators, and pumps to actually make a product move to follow the intended choreography.

Movement is often the most exciting of the three modalities we explore here in that it can affect the entire architecture of a product in dramatic ways, adding richer meaning to the stories that a product tells. A solar-powered lamppost with a head that rotates can give the impression of yearning as the lamp reaches up to face the sun. A security robot that patrols a street gives the impression of greater awareness of its surroundings than one that stays stationary on a corner. Indeed, any object that can autonomously navigate a space gives the impression of agency based on the fact that it can move its body on its own based on cues from its surroundings.

As compelling as movement is, it's challenging to develop because of the complexity of the electromechanical engineering required. Here are a few things to keep in mind when designing for movement:

- Motors require a relatively large amount of power to run.
- Motors require larger-capacity batteries.
- Products with movement need more sophisticated power management than products with moderate sound and light capabilities.
- Moving parts contribute to wear as surfaces provide friction against one another or stress certain parts of a form.
- Moving products need to be more robust than static ones.
- Moving parts can introduce some safety hazards, such as pinch points.

OBJECT LESSON

Clocky the Runaway Alarm Clock—"Catch Me if You Can!"

Clocky, made by Nanda Home, is a simple alarm clock on motorized wheels. When the alarm is triggered, it rolls away, off the nightstand and onto the floor, shrieking its highly annoying alarm until somebody chases it down across the room to hit its snooze or off button. For those with a tendency to abuse the snooze button or wake up so foggy they can't remember just why they wanted to wake up in the first place, Clocky can be the long-awaited solution to getting out of bed on time.

Clocky is not what we would call a "user-friendly" product, but it is effective. Its form boasts a rounded body, pronounced wheels, and an almost smiling face, which at first read seems friendly. After a few morning chases, however, the same form

FIGURE 4-5
Nanda Home's Clocky the Runaway Alarm Clock

---

reads as far more menacing. Its rolling is an exhortation, "Catch me if you can!" Clocky's ability to move is critical to its agency; because it does not sit and obey our sleepy orders, it can save us from ourselves in our weakest moments.

It is also not a particularly sophisticated product. Compared to app-based alarm clocks, it offers relatively few features. There's no option for multiple days or changes in display aesthetics. Nonetheless, it still feels like a roving robot. Because its motorized wheels move around in the space, it gives the powerful illusion that it's a creature moving of its own accord, thus triggering one's defenses to wonder what's invaded the bedroom, even if we can only see it peripherally.

Sound

In the ongoing search for efficient and engaging interactions, sound is a promising form of both feedback and input. Much the way my Italian cousin, Silvia, and I have a shorthand language that includes words, phrases, and fragments of sounds ("Uffa!" "Aiyee!" "Eh!") that represent a variety of ideas, so, too, can a person and a product share a similar lexicon.

In some sense, sound is the most natural way for people and products to conduct social interactions because it's at the foundation of how we communicate with one another as human beings. Whereas the semantics of light and motion require a translation from the original form into a message that we will decode (e.g., green lights indicate "all systems go"), sound can be delivered most directly to people in the form of verbal messages that we already know and understand. Whether they are language-based or warning sounds about what's happening in our environment (the footsteps of an intruder, the imminent fall of a branch, etc.), we have a keen awareness of the sounds around us, and our ears perk up to look for meaning in them.

Sound is invisible, yet it can be ever present throughout a space, traveling across a room and filling a volume. Unlike light or movement, it can be detected by a person who is not looking at the product and thus can offer a freedom to multitask and layer messages on top of visual or tactile signals. It can be used by an appliance, such as a washing machine or dishwasher, to indicate when a cycle is complete but can also be used to deliver highly specialized and sophisticated messages such as news stories or detailed stock reports.

Sound is also an extremely efficient means of communication if we consider how quickly a very compact message can be conveyed and comprehended. In *The Sonic Boom: How Sound Transforms the Way We Think, Feel, and Buy*, composer Joel Beckerman describes the phenomenon of rapid recognition that we've all experienced at some point in life when we hear a fraction of a musical note, yet the timbre of the music and the pitch of the sound combine to evoke a memory of an entire song: "When the songs are more familiar, then a lot of these motor or so-called premotor brain areas became engaged. Within milliseconds, you're not only recognizing the tune and feeling emotional about it but also rehearsing what action to take to respond to it."[9]

Making Sense of Sound. The human ability to parse and decipher sounds is more sophisticated than we may realize. Much of the sound that happens in our lives is fairly sloppy—that is, it comes from many different directions in a home or street, it spills over from one space to another, and we likely think of a world full of noise pollution rather than an organized and detailed landscape. Nonetheless, the human brain is able to decipher and distinguish sounds to a high degree of resolution. We regularly participate in what psychologists call *selective attention*, or the "cocktail party effect."[10] When having a conversation in a loud room, we are able to make meaning of what we hear from the person speaking while filtering out the other sounds, letting our brains essentially classify them as "background noise." When we bring a new product into our homes, we train our ears to listen for the new sound and distinguish it from other sounds in the environment.

In considering this phenomenon, a product strategy can include both foreground sounds that need to catch one's attention and background or ambient sounds that can provide information if a person is actively listening but don't demand immediate focus. A device monitoring a hospital patient's heart rate, for example, can allow a nurse or caretaker to know the rate is within an acceptable range with an ambient sound that blends into the background but also alerts them when it goes beyond range with a louder, more strident sound that demands attention. What's particularly tricky for nurses in this situation is that they can become accustomed to the alert sounds to the point where those, too, blend into the background while creating noise pollution for patients and family members, so it can be important to plan the sound design carefully to avoid the "crying wolf" effect of alarm fatigue.[11]

We can think of sounds in terms of two classes of messages: literal and representational. Literal messages are actual words, phrases, sentences, paragraphs, and so on that emanate from a product. Representational aural messages take place in the forms of tones, melodies, blips, bloops, and more. They are the bell tones that you hear in a car interior when the driver or front seat passenger hasn't fastened his or her seatbelt or the beeping that emanates from your microwave oven when the cooking time has elapsed.

Sound is not an embellishment but a reflection of the heart and soul of the product and therefore must be given a good deal of attention during the design effort. Professional designers are increasingly turning to sound design specialists in order to invest an appropriate amount of effort into this essential aspect of interaction.

Spoken Words Communicate Directly. Though it's a relatively new phenomenon for products to speak to us with language, it has become a nearly ubiquitous feature in contemporary consumer products, from the headphones that let us know they are connected to Bluetooth—"Connected!"—to the desktop speaker that announces the weather when prompted. Chatty products are beginning to find their way into every aspect of life, such as home safety, sports, security, entertainment, and medical devices.

Spoken words command our attention. They are also very reliable because they do not need to be decoded or translated. They feel natural because they are in the same language we may use to command the product to do what we need. When we say, "Alexa, what's the weather today?," it seems logical for the device to respond in kind, using the same language. Furthermore, we don't need to learn anything about an interface to interpret a spoken word message; simply knowing how to communicate as a human being is all the training we need to know how to use the product's interface.

On the other hand, do we really need all those words? We are living in an era of heavy language usage in new products that employ conversational agents and speak to us in full sentences and even add some unnecessary sentences of their own in the effort to render themselves to us. Every time I set a timer with my iPhone, for example, I hear the standard tones that let me know that my request was heard, followed by a voice confirmation that says, "Your timer is set for XX minutes, and the suspense is killing me." It's surely cute, but is it necessary? Similarly, when I ask Siri what the weather is, the device says, "It appears to be raining right now in New York, with a temperature of fifty-five degrees." While it can be comforting to hear sentences composed the same way a person might say them, the filler words like "It appears to be" are in many ways superfluous.

Dor Skuler of Intuition Robotics described a strategy around what I might call *digital integrity.* He explained a focus on transparency, including attributes such as a filter on the computer-generated voice to give it more of a robotic feeling, rather than trying to have the device mimic human speech. Led by Yves Bèhar's firm Fuseproject, the design team leaned heavily on multimodal "body" language through lights, sound, and movement instead of simulated facial expressions. He explained, "It's always an honest relationship that makes very clear what it is, what it isn't and what it can or can't do. It doesn't try to fool you."[12]

Sound as Opposed to Speech. As clear as verbal messages are, more efficient and perhaps even more compelling might be the development of an entirely different kind of language that the product uses to communicate with the person using it. And much in the way my cousin and I use a familiar shorthand with one another, people can develop relationships with their products that don't rely on full sentences or even full words but some new language or a hybrid of musical tones and language. Think of *Star Wars*' R2-D2 and how much personality he had without speaking a single English word. It can not only be efficient but can contribute to establishing a sense of character that's unique, memorable, trustworthy, and representative of the brand.

How can you develop this distinctive mode of communication? It's helpful to literally write out the conversation in "longhand" form, as if it were a screenplay. While the intention may not be to have the product literally say the phrases you've written, it's a great starting point for what is essentially a translation process. Your team can then place relative value on different parts of the interaction to determine which should be front and center and which may be more appropriate for the background of the user experience. Animator Doug Dooley describes this process in terms of the intonation an actor makes with the phrasing that travels up or down in pitch, with "up" at the end of the phrase being associated with more positive sentiments and "down" at the end of the phrase as being more negative.[13]

In addition, there are many practical and technical elements to keep in mind when crafting the sound output:

- *What's the quality of the speakers and the processor?* Not all systems are alike, and many times, due to cost or internal component size, the quality of the system may be relatively low. In this case it's essential to craft the sounds appropriately. Certain low-cost systems use sound that's generated by a chip and commonly referred to as eight-bit sound. It is limited in terms of its range and timbre, but if it's a known restriction of the system, then sounds can be composed specifically for it.

- *What does your brand sound like?* Just as an icon or a color palette represents your brand, so does sound. *Timbre* is a word used to describe the nature of the sound, and it will have a big impact on how your product is perceived. Imagine the difference between a violin melody and a banjo lick, for example. The former is reminiscent of classical music and formal theater halls, whereas the latter may conjure up images of hayrides and overalls. If there are verbal messages, those, too, should be crafted to match the personality of the brand.

- *How far is your product from the person?* Will it be important to communicate from across the room or in another part of the house? The end of a cycle for a washing machine, for example, is something you'll want to know about even if you are in a different room in the home. Therefore, the volume and pitch should be such that it can travel a long distance as well as be distinguished from the droning of the machine itself.
- *What are the other sounds that are likely to be in the environment?* Will this sound be competing with music? Will it be interrupting dinner party conversation?
- *What's the tone of the context?* An interactive yoga mat might need ways to communicate its state or offer feedback to the person using it, but given the environment and mood, sound may be disruptive and take the person out of a meditative state. In this case a muted sound may be best, if any.
- *How critical is the message?* The Nest smoke detector gives off a clear message when smoke is first detected: "Head's up, there's smoke in the bedroom." It also includes an important piece of information regarding the location of

the alert. Since this may be the start of a critical and dangerous situation, it's important that the message be loud and clear, though it's also best if it's not overly alarming so that a person can make a calm assessment of the situation. In some cases, escalating alerts may be appropriate so that the tone of the message, the pitch of the sound, and/or the volume becomes more intense as the situation becomes more serious.

OBJECT LESSON
Jawbone Jambox Bluetooth Speaker—"I'm Connected!"

Minimal physical outputs enable most of the Jambox's expression to take place through a series of sounds as well as spoken words. While there are buttons in the form of a chunky circle along with plus and minus signs, the interface doesn't depend on a screen output for feedback. Instead, small blips confirm button presses, and spoken phrases like, "Jambox is connected!" alert people to events such as low power and Bluetooth pairing. It has a signature sound and voice but also allows customization with voice personality and language preferences. For a brand that has built its reputation on high-quality audio products such as fashionable Bluetooth headsets, it makes sense to demonstrate the core value of the product—its sound output—through every moment of interaction.

Combining Modalities

When planning a product, it's essential to detail the ways that each of the modalities—light, movement, and sound—will operate in isolation in order to develop assets and assess overall messages. However, the experience a person has with a product will ultimately be created by a combination of all the dynamic characteristics at once. Below, I lay out the benefits of each as a guideline to deciding when to use each one.

Light

- *Localizable:* Whereas sound will interrupt a person by making its presence known throughout a space, light stays in the general area of the product, making it useful for communication that only needs to be noticed in context. The light on your rechargeable drill, for example, doesn't need to demand your attention when you're at the breakfast table.
- *Holistic:* Light can create a glow that fills an entire form, changing its overall presence based on its color and intensity.
- *Persistent:* A light that alerts someone to a message can remain lit until the message is retrieved. Sound and movement, on the other hand, are transient—once the message is delivered, it's gone until it's repeated again. Persistence makes it useful for messages that can be asynchronous. In other words, light is an effective use of a message that can wait for the viewer to pay attention when

he or she is ready. It's a little more passive than sound in that way and less startling than movement in that it can glow slowly and gently, appearing in a person's peripheral vision perhaps without their realizing its change in state.

- *Flexible in resolution:* Light can offer a block of color indicating something is on or off, or present or not, but the same product may also provide a matrix of pixels that form icons, words, and animations.
- *Glanceable:* In interior spaces, light can communicate simple messages well at a distance, so it's useful for alerting people to changes in state, such as an oven that's completed its preheating cycle.

Movement

- *Peripheral:* You can perceive movement from the corner of your eye and understand what a product is communicating without having to focus on it, so you can know what Clocky the alarm clock is saying while wrestling with the last minutes of slumber.
- *Rich:* Using movement can make an object more visceral than its still counterpart since it involves a physical, three-dimensional transformation, so knowing that Clocky is roaming around your room gives you more incentive to get out of bed, while the Polycam Eagle Eye offers the reassurance that it's no longer looking your way when it's offline.
- *Contextual:* The same movement can serve different purposes depending on the situation so that a desktop

robot's nodding could allow a videoconference to say yes or show you that the product's camera is scanning from the top to the bottom of an object.

- *Universal:* Since movement can communicate messages nonverbally, it can be used as a substitute or to enhance more verbal screen- or voice-based messages. Clocky's movement would be understood as a challenge to chase it regardless of what part of the world you live in.

Sound

- *Grabs immediate attention:* As opposed to being an ambient characteristic such as lights or movement that is seen peripherally, sound can pull a person's attention away from what they are currently engaged in, thus making it a great modality to use when a product needs to be engaged with right away, such as a seat belt alert.
- *Cognitively efficient:* Since we can process sound immediately and can associate meaning with very short sounds, it can be communicated in a fraction of a second.
- *Mechanically efficient:* Unlike motion, which usually requires the energy-hungry and space-demanding resource of motors, sound can be generated with a circuit board and small speaker.
- *Offers deep emotional associations:* Sound has the power to both offer the nostalgia of positive memories as well as establish new significance for the person using it and thus bolster a product's brand.

Up Next: Interaction

In the next chapter, we will move to the next ring on our framework, interaction, which takes into account what we know about product architecture and dynamic expression and then adds to the product the ability to truly interact by sensing the people and environments around it. The product then can not only offer messages about its internal state but communicate real-time responses as well.

IN THE LAB

Using Sound

When working for the firm Smart Design, designing a floor-cleaning robot for a company called Neato Robotics, I asked the team to break down the product's behavior into critical "moments" for personality definition. With people, we know that a person's true character is revealed during moments of extremes—both negative (stress, anger, fear) and positive (pride, jubilation, satisfaction)—and so we used these moments as key elements to define the robot's character. To create an abstract language of sounds, we detailed every message that the robot might have to convey in "human" language of words and phrases such as, "I'm done cleaning now!," "My battery is running low," or "Help! I'm stuck under the couch." Types of sounds were categorized so that expressive moments such as wake up or cleaning complete could take place through melodies, but other sounds could be expressed as alerts or feedback blips. Just like we might work with a color or material specialist in a traditional product, here we worked with a music composer named Skooby Laposky to create a palette of sounds. The human language was translated into a language of musical phrases and tones that conveyed not only the content of the message but also the emotional aspect of it, from the distressful alerts to the jubilant celebrations.

When the unit is powered up, it has a "wake up" sound that is also the signature sound for the brand. It's a short melody that's memorable and catchy but takes place in under two seconds. When it's beginning its cleaning cycle, it sounds another short melody, communicating the message, "I'm off to work!" A final melody takes place when the unit is back in its base and about to turn off, or go to "sleep." During the cleaning cycle, there are several potential moments of interaction that relate to both core cleaning tasks and social exchanges with the person. The moments were then earmarked for three different types of sound that would vary in tone, duration, and volume, depending on the context:

- *Melodies:* These were short musical phrases of a few notes each to signal a major change of state, serving the purpose of celebrating the robot's wake, cleaning, and sleep cycles. They could be heard from across the room but were not programmed to be at the loudest volume since their message is not critical to the unit's performance. When thinking about a product's expression, I like to identify the product's emotional state, so to speak. In this instance, the robot is jubilant, announcing either excitement about beginning the job or contentment at a job that's been completed.

- *Alerts:* These represented an announcement of issues or a call for user input. In terms of emotional state, they were cries for help or an indication that the robot was in a state of distress, such as when it was stuck under a piece of furniture, had an obstruction in its intake area, or was running out of power. These were extremely short sounds elapsing over a fraction of a second, usually containing two or three notes and expressing more negative emotions. Since they required user input, we had to consider the fact that the user might not be in the same room as the device, so they had to be loud enough to be heard at a distance. On the other hand, we knew that repeated sounds can quickly become annoying, so we tried to reserve these only for critical situations.
- *Input feedback:* These were the shortest of the sounds, intended to give the person confirmation that input was received. Rather than being musical phrases or even messages, these were the closest to short impersonal beeps of all the sounds. Since we knew that the person would be hearing these when pressing buttons on the unit and therefore was within a few feet, we knew they should not be programmed to be very loud.

5

Interaction Intelligence

The Rich Conversation between Objects and People

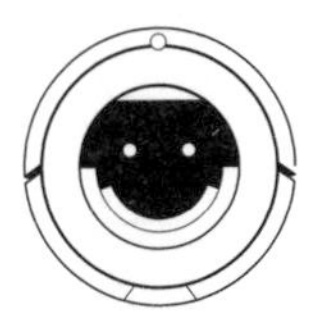

Meeting Simon the robot for the first time was one of those unforgettable moments. I was at the Conference on Human Factors in Computing Systems (CHI) in Atlanta in 2010, and it was at the first demo where I saw Simon up and running and performing interactive tasks with people.[1] As I had passionately toiled away for months on the robot's design details, I knew exactly what to expect. Essentially a hunk of metal, plastic, and electronic components, Simon had a microphone in its head, cameras for eyes, and sensor pads in its hands that would be triggered like little

Simon, the Robot that Relies on Social Cues

switches when an object was placed in them. It would decode commands, analyze pixels, and play out choreographed movements. "Hi Carla!" exclaimed Maya, one of the graduate students working on the project at the time. "He's programmed to understand a few commands, and you can ask him to do a sorting exercise with this green bottle."

"Simon, take this," I said, following the script. The robot extended its arm, and I gently pressed the side of the soap bottle into the outstretched fingers of its mechanical hands. It gracefully wrapped its fingers around the bottle. "Where does it go?" I asked, and what happened next was nothing short of magical. Simon held the bottle up to its eyes to study. In an instant, the lights in its antennae/ears lit up and glowed in the same green color as the bottle. Then Simon looked right at me and said, "It goes in the green bin." Though I understood the technical details of how the robot was programmed to parse the relevant words of my spoken sentences and respond with programmed speech and movement, I somehow still became lost in the moment, completely engaged with this marvelous creature. "He's looking at me!" I thought. "The green bin! He even knows what I'm thinking!"

It gave me goose bumps.

Through this remarkable exchange, I knew that my life as a designer had changed forever, as this sort of intimate and intuitive connection became for me the gold standard for what interaction should feel like. The fact that I, as a person who understood what was happening "under the hood," could be wrapped up in the illusion of life, made me understand without a doubt how powerful the feeling could be, and I began to reflect on a design approach for creating this effect in developing other products, one that leapfrogged the cognitive burden of complex interaction and having to "learn the code" of using an interface.

While it's important for product creators to understand that interaction can only take place through a collection of discrete, preprogrammed rule-based transactions between a person and a device—for example, sensor sees person, light gets triggered—the overall feeling and nuances of the experience should be seen in a social context. As consumers and as designers, we've been brought up in a world that thinks of a product as a collection of features, like *Bluetooth connectivity*, *3X optical zoom*, and *five-button*

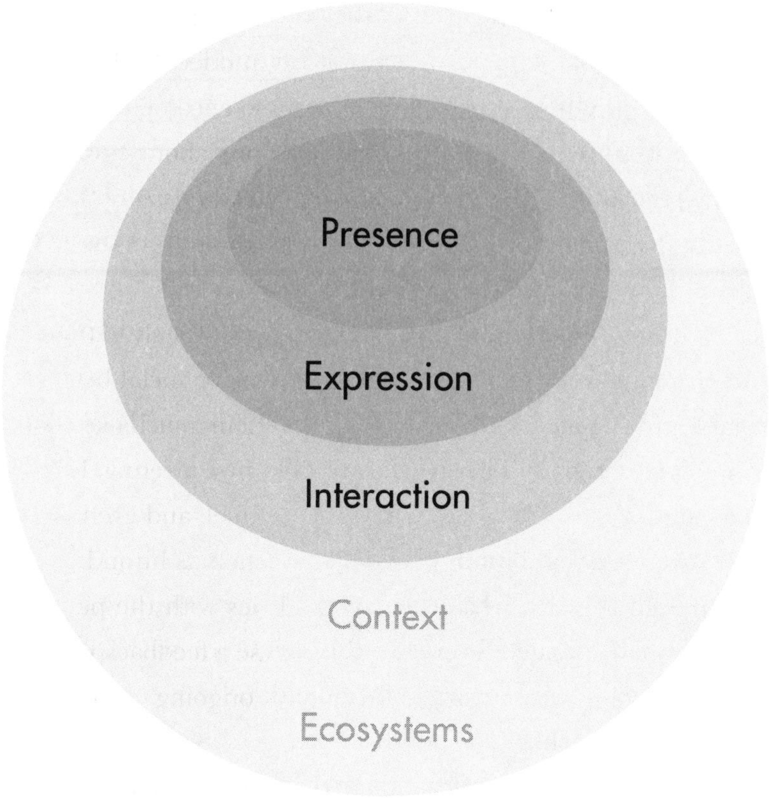

Presence
Expression
Interaction
Context
Ecosystems

FIGURE 5-1
Interaction, the Third Ring in the Social Life of Products Framework

control, but our best decisions will come from abandoning this approach in favor of a more holistic view of the overall relationship that takes place through exchanges between a person, a product, and its surroundings and then building a design plan accordingly. No matter how complex the product or device, the story of this social relationship can serve as a central, orienting strategy for design decisions during development.

Interaction is the back and forth that people have with the product. Building on a product's presence and its capacity for expressing messages, we now go one step further to consider a continuous ongoing exchange of messages and feedback between the person and the product. Interaction also encompasses activities that help the product to make decisions about automated responses, which are informed by *sensing* and *inference*. We can think about this as the device getting information from the world, rather than from the user.

The idea is to have the products react appropriately to the user and the context and thus behave as responsive social beings in active conversations with the people using them, much like Simon was actively conversing with me during our first meeting. It's possible to design products that can hear, see, feel, and even recognize the presence of invisible substances such as humidity and carbon monoxide. Combining these abilities with the power of expression gets particularly exciting because a feedback loop between those inputs and outputs creates an ongoing, evolving dialogue between the object and a person.

Though Simon is a highly specialized one-off machine created for laboratory research, the idea of interaction strategies based on a deliberately crafted social relationship is one that can be applied to many products with robotic capabilities.

Consider autonomous vehicles. Researcher Frank O. Flemisch and his colleagues introduced the idea of driving as a partnership in which the person and the vehicle share control. Imagine the relationship between a horse and rider. The rider maintains basic control. She can loosen or tighten the reins to vary the degree to which that control is exerted, but all the while the horse is still able to walk and make decisions on its own. They call this the *H-metaphor* to give a shorthand label to the concept of the overarching relationship—and not the technical function of the sensors and mechanics—as the guiding principle for the system.[2]

The heart and soul of a product's social life occurs through interaction, and in this chapter we'll look at the ways that interaction can be strategized, planned, and created.

The Basics of Interaction: Sensors and Actuators

Before delving deeply into interactions themselves, we need to review the nuts and bolts that drive interactive behaviors to ensure that visions are built on realistic possibilities for product design.

Let's look at an interactive system in its simplest sense. When you push a button on a machine and an LED lights up bright green in response, the button acts as the input, or sensor, and the LED response is the output, or actuator. Most systems, of course, are more complex than a button and a light, but the basic premise is one on which everything else is built—that is, the ultimate interaction will consist of a series of sensors and actuators with feedback that takes place in between to guide parts of the system. The feedback may come from a person, the environment, or some other part of the whole. That LED light, for example, might be a first-level response to indicate that the button pushing is registered and that a desired task, like brewing coffee, is underway.

The chapter on expression took a deep dive into what were essentially actuators. We looked at messages that products send outward in the form of light, sound, and movement, and the elements in a product that allow those messages to be expressed are in the form of specific components such as lights, speakers, motors, gauges, text displays, and so on.

The true "conversation" between person and product can only take place when sensing is introduced into the system to mediate what the actuator expresses and to constantly monitor and change the messages and tasks accordingly. Sensing therefore plays a huge role in streamlined interactions, so it's helpful to have a clear understanding of the potential elements of the designer's palette when thinking about taking in information through sensing systems.

Examples of Sensors and Actuators

Inputs—Values Read by Sensors

Pressing a button

Turning a dial

Swiping across a trackpad

Opening a drawer

Touching the base of a lamp

Stepping on a platform

Waving your hands

Speaking

Outputs—Values Expressed through Actuators

Little screens

Blinking lights

Recorded voices

Lights turning on

Sprinklers going off

Music blasting

The Nuts and Bolts of Sensing

From a philosophical point of view, it's hard for us as humans to truly understand anyone (or anything) else's experience, so we imagine it in the context of what we ourselves have experienced. For example, we envision that a dog enjoys food or hears sounds the way we do when in fact its experience is completely different, with certain senses, such as smell or hearing, amplified compared to ours and others, such as viewing colors, diminished. Ultimately, the essence of sensing for a dog is different in a way we can't possibly know. The same is true for products. Nevertheless, using human experience as our paradigm is a useful model in describing how we intend our products to decode the actions from the people and environment around them. Let's look at each of the human senses: touch, hearing, sight, taste, and smell.

Touch. Touch is the most common way we interact with products. From a simple, old-fashioned push button to the complex array of touch sensors on a smartphone screen, we have very well-established mental models for using our hands and fingertips to communicate with the objects around us. The most basic signal we can send to an object is the result of a switch that completes a circuit—envision two wires that are connected or separated. When creating a product, we craft this on-off value as a social interaction by taking into account the context of the engagement. A light switch gives us the power to turn night into day. It's located on a wall within the room where the light is

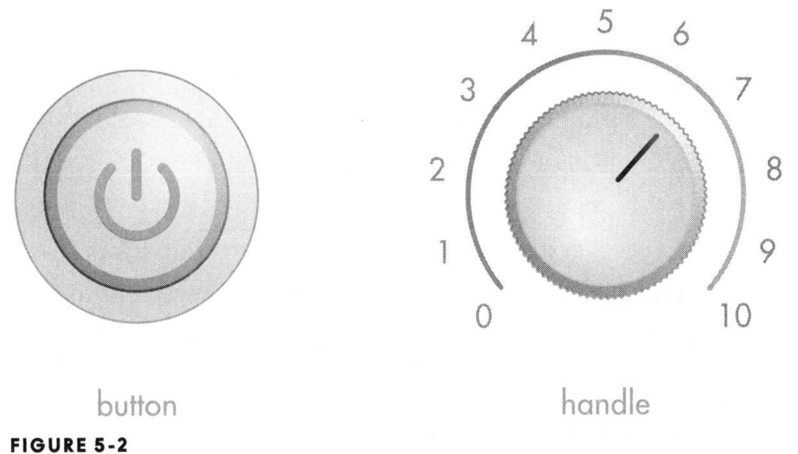

FIGURE 5-2
Buttons for Discrete Control and Handles for Continuous Control

located; flipping up, toward the sky, in the direction of a sunrise, brings us light; flipping down cloaks us in darkness. Though we take for granted seemingly mundane everyday interactions, they are impressive actors that connect us with our constructed environment, giving us a comforting sense of control.

Button or Handle? When designing a basic actuator, a key characteristic to decide upon is whether the person will need to communicate an on-off value, such as the simple switch (think *0 or 1*) or if the input will need to indicate a range, such as volume level on a stereo (think *on a scale of 0 to 100 percent*). The difference has implications both for selecting components as well as for envisioning how a person will interact. Design researcher Bill Verplank, who is credited in academic circles as being the father of interaction design, likes to describe these two as *buttons* or *handles*, where buttons refer to discrete control, like the keys on a piano, whereas handles offer continuous control, such as the slide on a trombone.[3] And although there are many standard ways of communicating both of these, what matters most for the social value of the product is how the gesture used to create the input relates to the overall situation. For example, it's appropriate to wake up a bathroom scale by tapping a foot on it since standing on it is the main way a person will relate to the product, and ergonomically, it makes sense to create a button that doesn't require someone to bend down to interact with it. The Nest thermostat amplifies the importance of setting the temperature in the home by making the entire object serve as a continuous control; the main body is encased within a circular ring that can be rotated to indicate the desired temperature.

Digital or virtual interactions are becoming increasingly tactile, and common design layouts capitalize on established mental models around buttons and handles. Most touch screens today rely on capacitive sensing: they detect when an electrostatic field has been interrupted, such as when a finger or palm is touching a nearby surface. By arranging capacitive sensors next to one another, complex gestures such as swipes and touches with multiple fingers can be understood. When combined with vivid graphics and animations beneath the sensing surface, they offer the compelling illusion of buttons being depressed, sliders being adjusted, and knobs being turned, thus expanding upon existing mental models from the physical world. Apple's Garage Band app capitalizes on these mental models through its series of virtual instruments on the screen.

Detecting movement is another aspect of touch that can be harnessed to create meaningful interactions for products. A tilt sensor is a simple device that has a metal ball that moves around in a cylindrical chamber. When the sensor is in an orientation that sends the ball to the bottom of the cylinder, it makes the connection. Technically, it is yet another way to create a simple switch: the equivalent of a very boring set of wires touching one another. However, the tilt switch allows designers to create magical interactions by associating the act of physically moving a shape with a predetermined reaction. One of my lamp designs, for example, is in the form of a wedge in which the entire object serves as the switch. When the triangle rests on its longer side, the light is off; on its shorter side, it's on. When exhibited at the Museum of Arts and Design, people returned repeatedly to the exhibit to play with the object as if it were an interactive building block.

This type of sensor can also be used to read shaking gestures. For example, if a designer wanted to create a music mixer that was activated by making a cocktail-shaking gesture, she could use a tilt switch to monitor the number of switch activations over time and translate those to understand the gesture. Similarly, small sensors can be used to create custom systems appropriate to both the interaction at hand and the parameters of the object's architecture. A coaster that glows may use the weight of the glass to connect two internal components to show a server when a glass needs to be refilled. A bird feeder that keeps photo records of feedings can be activated whenever the perch has weight on it. A water bottle can track its use with a sensor that knows when someone has tilted it upward to take a drink; it may even use the water as a conductive element that completes a circuit that tracks when it's empty or full.

The world of sensors around touch and movement is vast and can warrant an entire book on its own. The important thing to remember is that sensors can be used in creative ways well beyond the predictable pushing, pulling, twisting, and sliding to encourage gestural input involving gestures that are meaningful to the task at hand.

Hearing. In the 1980s few product commercials captured the public imagination as well as those for the Clapper, a device that allowed people to turn any appliance on or off by the sound of a clap. At the time the idea of controlling a device with anything but a switch or a dial seemed magical, and the promise of hands-free operation felt futuristic. The campy nature of the commercials also emphasized the improbability of this interaction as a necessary feature, and it would be forever remembered in product design history as a novelty that couldn't penetrate everyday life.

A few decades later, sound input has become commonplace, with devices such as the Amazon Echo, Google Home, and Apple Homekit enabling voice interaction to provide search capability, control media, and enable scan devices. We can certainly clap to control our devices, but with today's speech recognition software, we can also command, plead with, or cajole our products into doing what we wish.

Voice control has become an ideal way to interact with a device on a personal level where there is visual as well as audio output present, such as dictating an email or text message. And as directional microphone components become more common and affordable—the Amazon Echo has eight microphones situated in a ring around the perimeter of the top of the device—controlling the objects and media in our indoor environment is becoming a much more desirable way to interact with our products.[4]

There are many applications that call for voice control as the best possible method of input. Consider situations where we need our hands free to manage the task underway, such as cooking a messy recipe or holding a baby. Being able to say, "Alexa, set a timer for twenty minutes" or "Call Grandma" lets people quickly do things that previously required manipulating a device. Being able to dictate also allows commands and messaging to take place in situations where it's not practical to look at an interface, such as when driving or walking.

Using voice also has many limitations and is best used in situations that are personal and/or indoors. Trying to control objects on the street, such as a traffic light or items in a bus kiosk, could end up frustrating the person using it since it's hard to target voice in an outdoor environment competing with all the ambient sound present. Similarly, in a situation where many people are present, such as at a trade show or supermarket aisle, voice control not only will be difficult to execute technologically but will be awkward socially; it's easy for the interface to misunderstand who is talking and difficult for the person to know who has the interface's attention if there is more than one person in the area.

Sight. From night-lights to drones, many products rely on sight as the main way that input is received. While cameras may be applicable for some situations, they are often overkill for the level of input that the product needs. In general, the simplest—and perhaps crudest—sensor that can detect the input quickly and reliably is the best choice for the design, avoiding the processing power demands that may come from having a more complex sensor in place.

For example, a photocell is perhaps one of the most basic and inexpensive ways that we give entities sight. It's a small, light, and inexpensive component that changes its resistance based on the amount of light that hits it. This measure of resistance can be used to make decisions in the programming of a circuit (for example, "If the resistance is below one hundred ohms, trigger a sound file to play"). Solar-powered garden lamps that turn on automatically when it's dark out utilize these sensors, thus conserving energy by turning off when not needed.

On the other end of the spectrum in terms of complexity is using a camera as sensor. Instead of reading just one point of light or one color, a camera input enables the object to read a matrix of values that can be interpreted in different ways to learn a great deal about the environment or the user. Cameras can be used to understand a product's environment, so they can autonomously navigate a person's home or office, as a security robot does. They can also help in monitoring processes that are in progress, such as the Makerbot three-dimensional printer, which has a camera that's trained on the print bed; it can detect common issues such as a lack of plastic material due to the filament being stuck or having run out; it can also be used to communicate with the person using it in a remote location so that they can verify an issue and stop the print or approve its continuation. There are many new camera-enabled products that replace traditional doorbells to give people X-ray vision to see who's on the other side of their door as well as inspect packages or papers that might have been left on their doorsteps.

Camera-enhanced products can also give us the option to use gestures to communicate with our products. The combination of hearing and sight would allow an interaction such as the following to take place: Imagine a robotic vacuum named Oscar sitting in its charging base in the corner of a living room. Lucille is entertaining guests when she reaches across the coffee table and overturns a bowl of almonds, which tumble to the floor. "Oscar, can you come this way?" As the robot moves toward her, she points to the mess on the carpet. "Please vacuum this." The interaction can take place smoothly with only a limited interruption of the party.

There are also examples of products that go beyond the conventional paradigm of human sensing. For example, the smart mirror from the Massachusetts Institute of Technology (MIT) Media Lab Affective Computing group can read a person's heart rate just by having them sit in front of it. It looks at an image of the face and then detects variations in skin color to interpret heart activity.[5]

The Camera as "Everything Sensor"

Until recently, the use of a camera embedded in a product would have been cost-prohibitive, but now we can consider this to be a viable product configuration. In fact, it is quickly becoming the most versatile and ubiquitous sensor, allowing a product to see and map out its environment as well as understand gestural input and read human conditions such as emergency situations and medical conditions.

Researchers at MIT's Computer Science and Artificial Intelligence Laboratory (CSAIL) have been developing a system that can manipulate video footage in order to exaggerate details and reveal conditions that would otherwise have been essentially invisible, a technology that's already finding its way into products from home to environmental applications.[6] For example, the Miku Smart Baby Monitor can track changes in heart activity and breathing without ever coming into physical contact with the child and uses infrared technology to function in the dark. Instead of thinking of a product that is sensor enabled, we can think of a sentient environment, such as a gesture-driven office or media room.

There are, of course, many downsides to using cameras as sensors. While inexpensive, their cost is still higher than simpler components. The processing power required to understand their data is demanding, thus increasing the complexity and expense of the entire product. And perhaps the biggest drawback that's also the thorniest to resolve is the fact that cameras present a serious compromise to personal privacy. Like microphones, they often need to be constantly on, and for some features to work well, they may need to record audio and video.

While there is great debate among manufacturers about the ethical use of cameras in everyday products, designers can shoulder some of the responsibility by considering ways that a camera's presence and status (e.g., recording versus not) can be made apparent and controllable. As customers become more savvy to the ways their data is being stored and used, making camera data use apparent and understandable will present an opportunity for product creators to shine in this area. Experiments by IKEA Space 10 researcher Timi Oyedeji showcase prototypes that let people control elements like lighting, curtains, and audio systems with sophisticated hand movements and other gestures that can seem like a natural extension of language. He is exploring using radar-based alternatives to cameras to preserve privacy.[7]

Multiple Senses at Once: Zero Interface?

I've been discussing the capabilities of individual sensors, but the most powerful systems utilize a combination of sensing techniques to develop smooth and responsive social interactions that more resemble a holistic experience, approaching the way people actually interact with one another.

Autonomous vehicles don't rely on just one type of sensor but rather combine radar, lidar, camera vision, and physical sensing to understand the intentions of the human riders, as well as those of the riders in other vehicles and the pedestrians who will cross their paths.

Amazon GO, the in-person shopping experience with no lines or checkout, calls its system *sensor fusion*. Combining sensors to understand a person's movement creates a "frictionless" process in which a person can just walk into the store, place items in a bag, and leave. From the Amazon GO marketing description: "So how does it work? We used computer vision, deep learning algorithms and sensor fusion, much like you'd find in self-driving cars. Once you've got everything you want, you can just go. When you leave, our 'Just Walk Out' technology adds up your virtual cart and charges your Amazon account. Your receipt is sent straight to the app."[8]

There is a movement afoot in the design world called *Zero UI*, whose proponents argue that people shouldn't have to think about controlling machines at all, similar to the Amazon GO experience.[9] When you walk into a room, the light should come on, when you shut the door to the dishwasher, it should begin washing any dishes that are inside. In these cases the input may be invisible to the user, but there needs to be a motion sensor to let the computer know that a person came into the room, or a button and perhaps a weight sensor to help the dishwasher to know when it should activate, and so on. It's hardly zero interface, but the cognitive burden is taken off the user and taken over by the system.

These are *implicit inputs*, put into place by the designer, who has some model of the way the machine will be used. These implicit inputs take the effort of "operating the machine" off of the user, but because the machine is *inferring* commands or calls to action, they also open up possibilities of error if the model the designer has built the system around is incomplete. This is where social intelligence becomes most salient. It might seem like a contradiction that products that use sensors so users don't have to turn products on and off explicitly are usually billed as "intelligent or smart," but they function in a way that is most natural when done right. In other words, "Knowing when to interact is interacting well."[10]

Testing Things Out: Using Microcontrollers to Prototype

When a person pushes a button on a product, what happens beneath the surface is that the voltage on an input pin on a microcontroller is triggered. Digital inputs, like buttons, have states that are either on or off, and analog inputs, like handles, can have a range of values, like 0 to 100 percent. The machine can be programmed to respond accordingly by having actuators create aspects of expression like playing a sound, waving a flag, or flashing a light. As opposed to earlier, more cumbersome electromechanical systems, manipulating the correspondence of input and output in microcontroller-based devices is just a matter of programming.

While aesthetics, and certainly the overall presence, of a product are core to its existence, designing an interaction without being able to try it out is like showing a kid a picture of a bicycle and expecting him to understand how to take off with it. Great interaction design emerges from having a visceral understanding of what an interaction might feel like, even if that interaction is cobbled together from bits of cardboard and clay with sensors duct taped to the inside. When I use a radio-frequency identification (RFID) tag (those pesky stickers on products that make the thresholds in stores beep) in a design, for example, I have a feel for just how far a scanner needs to be to detect it as well as how well it can be hidden inside a form and how it needs to be positioned to be effective. This kind of insight comes from trying things out.

Just a decade or so ago, designers never dreamed of creating interactive prototypes. The investment in time and resources was way too great and confined to engineering teams and research and development (R&D) labs, while empty or "dumb" design models represented the look and feel of a product. Today, a plethora of off-the-shelf and easy-to-use microcontroller platforms exist that allow anyone to create a crude prototype in just a few hours for under one hundred dollars or so. Open-source and online software and community forums offer step-by-step instructions that can be accessed just by searching for simple descriptions—*Cat-activated fan, can-opening robot*—so making a "works-like" prototype of something like an interactive file cabinet is within reach of anyone on a product team, not just the hard-core engineers.

The best platform for starting out with basic microcontroller prototyping is the Arduino, a board that costs about twenty dollars and consists of two rows of plastic into which wires can be inserted. Those wires can be connected to simple sensors or actuators, and then the whole assembly can be connected to a computer via a USB cord in order to program what instructions should be followed. For example, to make a simple night-light, a photocell can read the ambient light, and the code can look for a value, such as 80 percent dark, that needs to be reached to turn an LED on. Arduino code starts with very basic systems, like the night-light, and progresses all the way to complex robotics with

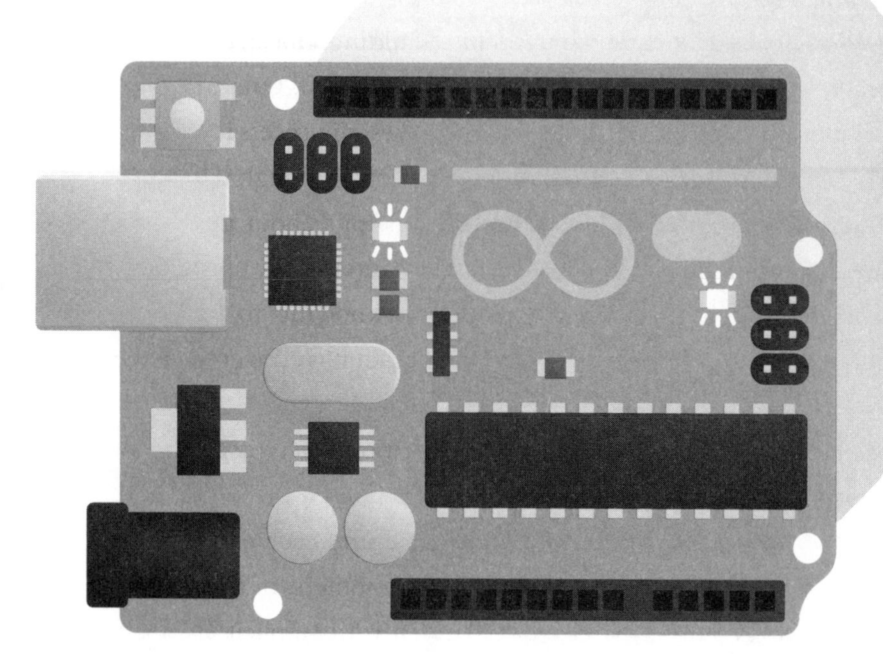

FIGURE 5-3
The Arduino Board for Basic Microcontroller Prototyping

highly sensitive inputs and choreographed movements and sounds for outputs.

Another popular platform is the Raspberry PI, which operates as a minicomputer and uses the Linux operating system. It's a good choice for anything that might use a graphic display, need to access information from the internet, or require complex processing to operate. It also has an immense online community that has contributed volumes of sample projects that can be downloaded and modified to suit any project at hand.

Thinking about Social Interaction

Now that we've covered the sensory building blocks of interaction, let's move on to the social interaction itself. One approach to applying the social perspective to the design process is to translate everyday interactions into conversations. Though not necessarily conversational, like having discussions about grandkids and politics with your toaster, it's a social interaction that's transactional and pragmatic—for example, knowing to look at a person (or machine) before speaking so that it is clearer to whom you are talking.

For example, table 5-1 shows a translation of a typical interaction with a ringing phone.

By treating the phone as a social actor and each exchange of information as part of a conversation, the comedy of the phone's lack of sensitivity to your situation, and your implicit cues, are more apparent. Part of the challenge comes from the fact that you can't easily anticipate when the phone is going to ring or be ready for the ring at all times. Another part comes from the fact

TABLE 5-1

Translation of a typical interaction with a ringing phone

Interactant	What is happening	Implied message
Your phone	Ring ring!	You have a call
You	[Your hands are full, you can't pick up]	I'm occupied
Your phone	Ring ring	You still have a call
You	[Trying to cover phone with your body]	I'm busy, be quiet!
Your phone	Ring ring	You still have a call
You	[Glaring at phone]	I'm mad you're still ringing

that the phone has the affront to ring loudly, without any accommodation for the situation you're in, whether the volume is appropriate to the situation and whether you're ready or not to answer the phone. The final challenge comes from the fact that you have many ways of trying to shush the phone that it's insensitive to; it will accept either active answering or being ignored but no other input at all. Because we all use social intuition in our everyday interactions, thinking through the implicit conversation in our interactions makes it easier to see gaffes and opportunities that we might otherwise miss.

Again, the traditional way of thinking about machines is in terms of components, so of course it makes sense that the phone can either be answered or not, like a receiver that is placed on a cradle or picked up. Thinking about the world of sensor possibilities, however, opens many new opportunities for a device to respond appropriately as a social actor. With a camera-based sensor, for example, the phone could read something like a glaring expression or a "Shush, please!" voice command and then respond accordingly. This is the type of subtle interaction that we can begin to invent and employ as designers.

In looking back at our interaction models, it's helpful to think about whether or not it makes sense for the conversation to take place *through* the product or *with* the product in order to decide how to craft interaction.

How Do We Design Interaction?

The core structure of what's needed consists of mapping out the conversation that takes place between the person and the product, with requests or feedback from a person detected as inputs

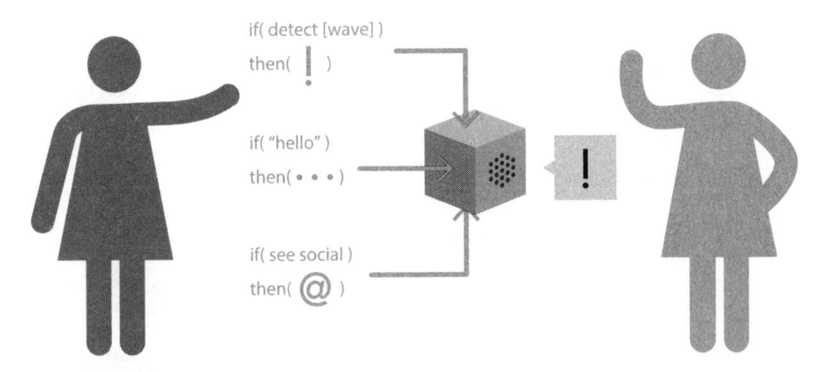

FIGURE 5-4

Interaction between the Designer and the Person, with the Product as Intermediary

The conversation here takes place between the designer and the person, with the product as intermediary. The language may even refer to the product in the third person, such as, "Remove the printer cartridge and inspect the ink."

and responses or feedback from the product as outputs. While the ultimate conversation may seem streamlined and simple, the effort that goes into determining what that dialogue should be involves a great deal of deep work, starting with design research, to truly understand what a person needs from a product in every conceivable use situation. For example, is messaging taking place through the device or with the device?

To begin this mapping, the fundamental story of the person's relationship to the product will serve as a cornerstone to developing a design strategy. Drawing on our mental model of the person-product relationship, it's helpful to establish how the social interaction will take place.

Once an approach based on overall story has been decided upon, there are still many open questions regarding how the social relationship will manifest in interaction details. To flesh these out, a script can be developed and then explored through a variety of design research methods.

FIGURE 5-5
Interaction between the Person and the Product

The conversation here takes place between the person and the product. The product may refer to itself: "Check the cartridges underneath my cover. . . . I'm ready."

While traditional marketing research would suggest a survey-type approach to gathering information for product development, this falls short of enabling a true understanding of the potential for products to serve the people using them. Offering only a shallow and reactionary point of view, surveys can serve designers in the initial stages of organizing thoughts around a product category or selecting research participants from a large pool but in general are not deep enough to inform design decisions.

This is where user research—or studying interaction in the real world—comes into play. There are many approaches that can yield richer insight that can drive a project forward. Some involve directly interviewing or *shadowing* a research participant or groups of participants. Others involve imagining experiences through drawings, videos, or playacted, imagined situations. And many are a combination of the two, in which proposed artifacts are used as probes or prompts for conversations to yield information that can inform the design process. Here is a list of some of these methods, a few of which I'll describe in more detail:

- Journey map for an overview of all aspects of the experience
- Identification of the key moments of interaction, mapping those moments to messages given and received by the product
- Good old-fashioned sketching
- Consideration of the relationship to the body: ten-inch experience, two-foot experience, ten-foot experience, remote experience
- Bodystorming interactions, key moments, and "paper" prototyping
- Wizard of Oz
- Animation and stop-motion animation
- Relationship arc: How will it change over time? (see "Long-Term Interaction")

- Multimodal flowchart
- Dialogue writing: planning the conversation

Enactment

In early stage design workshops with clients, I typically employ a variety of techniques that involve enactment, or having participants act out the behaviors of both the people and products that may take place during an interaction. Enactment allows designers to explore scenarios in real time and space without the limitations of current technology or of a prototype that has already been formed. It enables a focus on the dialogue and social interactions while leaving the implementation open for exploration and interpretation. Anyone on the team can participate in enactment techniques and benefit from the insights revealed.

For a quick idea of how enactment works, think of the game that you may have played as a child when you and three friends sat in chairs pretending to be in a car. One person may have held a frisbee to "drive" the car while the other three were passengers looking out the "windows" or conversing with the driver. A similar activity would be a great starting point for considering the experience inside an autonomous vehicle. The fact that it's free from physical walls or constraints can even lend itself to participants envisioning themselves in several vehicles that can communicate with one another or as both passengers in a vehicle and pedestrians that are passersby.

Often referred to as *bodystorming*—or embodied brainstorming—the technique is most commonly employed by having people perform the actions that a person using a product would perform. Ideally, the activity would take place in the context in which the product would be used, as well as with props made of simple, inexpensive, and malleable materials.[11] For example, if bodystorming an oven, one person might play the role of the person using the oven, and another person might act out the behaviors of the oven itself. (Yes, I've asked grown people to pretend to be an oven![12]) A cardboard frame held by the person representing the oven could represent the range top or door. The person playacting its use might approach it, indicate how he or she would touch the top of the console, or give the oven voice commands. The "oven" actor could then respond appropriately in real time, perhaps with abstract "spoken" tones or with phrases that could later be translated to appropriate behaviors for the appliance in that context.

The immediate benefit of these types of bodystorming exercises is that they reveal critical aspects of the product-user relationship that would not otherwise be obvious through more removed representations such as drawings, renderings, or scaled models. Working in life-size scale and in spaces that represent the final use (an oven bodystorm session in a kitchen, for example) can give deep and immediate insight into the situational constraints and opportunities. Additionally, focusing on the conversation between the object and the person rather than fixed product details encourages open-ended explorations that can move in unexpected directions. Desired product features might spontaneously emerge that can then be designed or explored further, and the technique lends itself to thinking about whole-body interactions, ideally freeing the designers from thinking only in terms of the more limited types of input architectures that have existed in products in the past.

Bodystorming techniques not only yield rich insights into social aspects of product design but are low-cost ways to rapidly work through ideas before investing resources in prototyping or locking down decisions about features and technology. In one study run by interaction researcher and Cornell Tech associate professor Wendy Ju, her team led participants in envisioning future autonomous driving experiences and asked them to describe how their time would be spent during this future experience. They shared the emotions this situation might bring and imagined the value that might come from relinquishing driving control and gaining back the time and attention usually lost during a commute. In addition to these open-ended sessions, they used the enactment exercises to probe specific use cases, such as investigating how drivers might take back control from partially autonomous cars at the end of the drive.

From such enactment sessions, the team was able to access and ideate around factors for trust and comfort, as well as ideas for new features, gestures, and movement in the car. They leveraged insights to inspire ideas for elements such as seats, steering wheels, music and climate controls, and interior details.[13]

Contextual Inquiry

While bodystorming that's free of any constraints that might be present with actual physical artifacts and prototypes can yield rich insights, it's also valuable to conduct interviews in the context that most closely resembles what would exist for the product that's being designed. When envisioning a piece of hospital equipment, for example, it is crucial to speak with nurses, doctors, and technicians on-site in the actual spaces in which they work. Talking about a design for a laptop cart while in a hallway might reveal ways that nurses need to move their bodies to pass quickly, and this observation could impact the form of the cart. Details such as the gesture of removing gloves before using a console might be revealed and addressed during this type of on-site conversation, whereas they might be missed if the interview takes place elsewhere.

This method is often called contextual inquiry, and in addition to being helpful for exploring current behaviors and experiences, it can also be used to help people envision future settings. When Wendy's team was considering how people might situate themselves in automated vehicles, interviews were held in a right-hand-drive Jeep Cherokee parked safely but with the engine running. Offering a setting that was as realistic as possible enabled the research participant to suspend disbelief. This enabled them to experience the uncanny feeling of being a passenger in a driver's seat devoid of operator controls. In this situation they could envision what an autonomous vehicle might do and describe how those actions felt.

Participants were also given props, such as pillows, phones, tablets, and computers, in order to probe into activities they might perform if freed from the responsibility of driving. One important insight that emerged from the exercise was that automation affords new social dynamics for passengers. If it is no longer necessary for the driver to face forward, then perhaps new social situations could be encouraged by adjusting seating to face away from or toward one another.[14]

Scaled Scenarios

Often with products that are large in scale compared to the human body, it's helpful to enact scaled-down scenarios by play-acting "from above" using props that are shrunken in size so they can be easily manipulated and viewed within the context of surrounding products or environments. It may feel like you're playing with cars and cardboard models the way a kindergartener would—and in some cases you're doing just that—but having a fun process doesn't diminish its value.

For example, when I was designing a breast biopsy tool, I used two-dimensional maps of the doctor's office to glean an overhead view of the dynamics of the space. Since there were critical relationships taking place among the patient, doctor, nurse, and technician that would affect the patient's comfort and the ability to successfully obtain a biopsy sample, having the map view helped facilitate team discussions about how permutations to the overall setup might improve the use of the device. These improvements, such as the technician's line of sight or the doctor's view of the sample area, could then guide specific explorations around the product's form, the tool's orientation, and interface details.

A three-dimensional scaled model can also be valuable in many social design contexts. In the case of automotive design, having scaled models of multiple vehicles that might encounter one another in a street scenario, along with models of street furniture such as curbs, benches, and parking meters, can yield systems-level thinking that can reveal solutions that span across products and services. For example, acting out the parking scenario in this case might lead to concepts around automated parking and metering and proposals to municipalities to retrofit existing meters to benefit from new systems that may eventually be implemented on a large scale.

In one of Wendy's projects, researchers reviewed video clips from studies that demonstrated patterns of behavior between drivers and between drivers and pedestrians. For example, when two parties are going to intersect in their trajectories, the faster-approaching party will likely make an adjustment further in advance to that intersecting point.

To analyze these patterns, she and her colleagues asked groups to improvise scenarios using diagrams and simple three-dimensional objects as props. Videotaping from directly above each station, they captured the movements and gestures that the participants made with their objects on the diagram, as well as anything they said that added to the scenario.

They ran through different interaction conditions, which included an oncoming interaction in the normal direction of travel, an oncoming interaction against the normal direction of travel, and a perpendicular interaction. After completing those structured but improvised scenarios, they asked the participants to come up with their own condition and design a scenario from that condition. This opportunity led to many creative interactions where the autonomous vehicles took on more emotive personalities and expressive human and animal behaviors.

Because the exercises were based on real, observed situations, they were grounded in real-world needs and tendencies; however, the open-ended use of the scaled representations encouraged innovative ideas and rich conversations around what interactions might be possible.

This project is also a good example of the idea that developing and implementing research methods can be a fluid process in which improvisation is embraced, and multiple exploration techniques can be combined and used concurrently.[15]

Wizard of Oz

Wizard of Oz (WoZ) is a powerful technique for studying a person's relationship to an interactive product by quickly creating

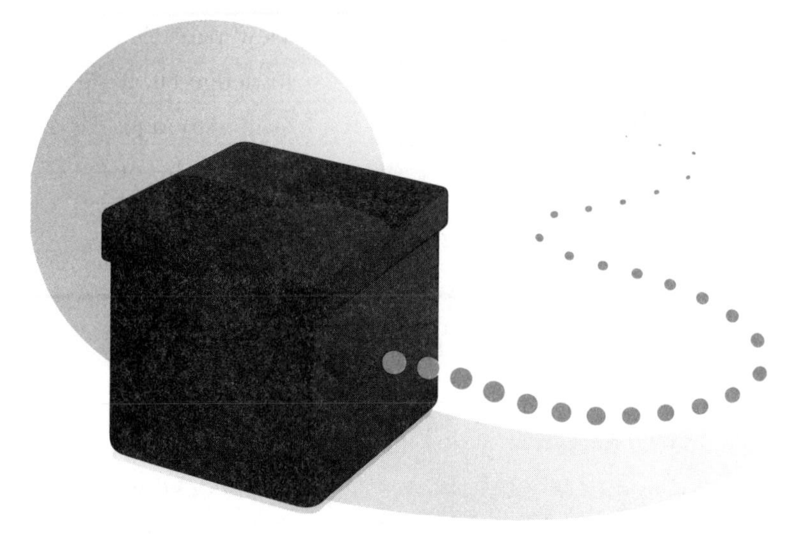

FIGURE 5-6
The Mechanical Ottoman Robotic Footstool

and implementing a dynamic experience in real time through cleverly positioned remote control, similar to the Wizard of Oz in the famous children's story. For example, if developing a chatbot, existing instant messenger software could be modified to present the interface as a conversational agent. A participant might begin a dialogue with the bot, thinking it to be software controlled, but instead the typed responses are generated by a person in another location.

At the Center for Design Research, Wendy and her team used WoZ techniques to explore how people interact with expressive movement in machines through a project they call the Mechanical Ottoman, a robotic footstool that can automatically drive around to position itself by a person's feet and pop its seat cushion lid up and down for an added element of expression.

In participant studies the ottoman approached unsuspecting people to see if they would understand the ottoman's invitations to engage or disengage. It would drive into view, pause, and wait at a respectful distance until the person acknowledged it. It would approach and lift its lid a little. Some people understood just what the ottoman was for and lifted their legs so that the ottoman could drive right up under their legs, and they could rest their feet on the seat. Others appeared a lot more tentative and required nudging to engage. If, after a few nudges, people did not lift their legs up, the ottoman would pop its lid a few times to indicate a desire to disengage, and then it would drive off to the next room.[16]

One woman looked a bit freaked out and screamed with surprise laughter when the ottoman approached. Several participants understood that the ottoman wanted them to put their feet up but felt uncomfortable doing so. "It seems like a dog, and I wouldn't put my feet on a dog. I wouldn't want to trap it." When the ottoman left, it was clear that many of the participants were a bit sad. One was even afraid that she had hurt its feelings somehow.

The Mechanical Ottoman prototype was driven by an iCreate robot—the same engine used in the Roomba vacuums—with a servomotor to pop the lid up and down. It was remotely controlled by a person, the Wizard, who was hidden in the next room. The Wizard watched the interactions via a web camera and remotely controlled the ottoman based on people's responses. This use of a Wizard in study interaction allowed them to adapt the ottoman's behavior in response to what each participant was doing; the Wizard's intuitions about appropriate social reactions are as much a part of the experiment as the behavior of the study participants. This kind of study let the team figure out what kinds of behaviors should be built into autonomous robots and what range of motions are needed to communicate.

Field Experiments

While a great deal can be gleaned from controlled studio experiments with design probes in place in an office setting, it's often most valuable to conduct research in the actual, real-world setting in which the product will be used. Unlike other techniques, such as video prototyping, which require some imagination on the part of the participant, field experiments can capitalize on natural interactions with a design probe, allowing a researcher to observe spontaneous, unplanned events offered by the real world. Participants can be asked to simply perform whatever tasks they normally do, and everyday experiences can be revealed.

For example, in the car interior project described above, my colleagues and I accompanied participants on typical driving routes that might take place in their lives on a typical day. Sitting in the back seat, we observed them running an errand at a local store, commuting to work, and picking kids up at school. Actually being in the car in traffic allowed us to see changes in behavior depending on the time of day, the errand at hand, and the unpredictable environmental conditions such as weather and traffic. We used this technique to observe participants' interaction with the car interior details that were already in place and then contrasted that with our design probes, which consisted of a touch screen dashboard interface and several other mock-ups, such as a simulated door panel made of foam core. This revealed many insights about the way they needed to interact with their interiors, including things that weren't necessarily "part" of the interface, such as how to share snacks with back seat passengers and how focus on the dashboard changes when weather conditions are challenging.

Video, Photo Sequence, and Animation Prototyping

The above techniques are helpful for observational research, which can be recorded and later reflected upon to develop insights. Bodystorming, for example, can be captured through a sequence of images that illustrate the interaction, or the overall conversation taking place between the person and the product. When developing the robot Moxi, for example, I facilitated workshops in which team members acted out the roles of the robot and the nurse, using real artifacts in a simulated environment (more on that in the "In the Lab" section). But before observation even takes place, prototypes of intended interactions can be created using visual means to simulate behaviors as a way of illustrating the use to potential users. For example, my studio has been developing a robotic lamp product that takes cues from people in order to change positions, illuminating a point on a desk or flooding a wall, depending on the circumstance. The proposed design involves voice commands and gestures to activate the lamp movement. As an initial step, our team built an articulated mock-up out of foam core and pins that could move through key joints in order to achieve the different modes we envisioned. The pins were situated where motors would be, and we used stop-motion animation techniques, taking one picture at a time, then changing positions, to approximate the intended motion without having to invest time into programming and building actual motor joints. With an accumulation of photos strung together, we were able to build believable interactions, staged along with a person to fully demonstrate the core conversation. The person could be seen gesturing in a motion that pointed from the desk to the wall, and then the lamp followed suit by changing position from a downward spot to a light pointing upward and from a focused light to a more diffuse one.

Physical Prototyping

Once a sufficiently "baked" script has been determined from the exploratory methods above, there will be key questions that continually emerge from the process of translating that script into the means of expression discussed earlier. For example, in designing a compartment for a delivery robot project, the team needed a way to test out the visual language for providing feedback for the compartment's state. "Should we do flashing red to indicate that the drawer has been closed but not been locked with the key card?" we wondered, along with the visual cues for the other states (fully open and closed and locked). This is a great stage for bringing in the off-the-shelf microcontroller kits, like the Arduino and Raspberry PI. Since the focus of the question is around the drawer, a quick mock-up can be created with foam core, and LEDs can be programmed to correspond to locking states. While something like a paper prototype of Post-its with colored marker drawings to indicate the lights might suffice in earlier stages in the process, here, having the real-time, visceral presence of the glowing light will offer a more direct and accurate sense of the people's perceptions in order to map out the precise timing, color, and intensity of the interaction.

Long-Term Interaction

All the interaction prototyping methods listed here, and essentially most of the existing methods, focus on new design scenarios—that is, the introduction of a new product interaction.

While this is essential, there is also a great deal of the core conversation between a person and product that will evolve over time, such that the input translations and output reactions make sense during the first few uses, or even first year of use, but may need to change as the relationship moves on and matures over time.

When designing the Neato Botvac interaction, for example, my team and I mapped out the product-user relationship over the span of time of ownership of the product. We laid out four main stages to the relationship, beginning with the out-of-the-box experience, which can best be thought of as a honeymoon period. Particularly with a product like a robot that is likely to be a novelty experience, we felt we could plan amped-up interactions to heighten the connection and exaggerate the product character. Lively sounds and expressive movements can be endearing upon first use, and they can also be capitalized upon to create a tutorial moment when the robot demonstrates its capabilities through its light, sound, and motion.

The second stage we envisioned was first cleaning, when the robot is no longer in the tutorial mode, so some of the more performative aspects of the interaction might be toned down. The relationship is still a new one, so it makes sense to portray key character traits through the way the robot behaves.

The final stage would be routine use, and just like any relationship, a product and the person using it will fall into a matured relationship in which the novelty has worn off, and the product may be best falling somewhere into the background of activities taking place in the home. In this case, some sounds would be toned down or eliminated, and other flourishes would be modified to call less attention to the robot unless necessary for maintenance.

Pick a Method, Learn from It, Repeat

Designing interaction is a daunting task, as it basically requires envisioning a new relationship between a person and product and trying to anticipate all the unknowns that will occur through use. There are so many methods that crafting the research plan is a project in itself; however, it's important to avoid being paralyzed by the wide-open space of unknown information by simply starting the research process anywhere and responding to insights as they emerge. In other words, to get started, simply start with *something*. That something will expose more questions that can guide what you'll design next.

With insights from the above methods, the kinds of interactions that might take place can be better defined, and the team can begin exploring details of the person-product exchange through specific moments, developing props and interactive prototypes to serve the more project-specific design questions that emerge.

INTERACTION TAKEAWAYS

✓ The person-product relationship will largely be defined by the conversation that can take place between them, so the designer's most important job is crafting the dialogue that will take place, with words and phrases replaced by inputs and outputs.

✓ Inputs and outputs can take a variety of forms and can be made up of many more shapes, gestures, and materials than have traditionally been built into products.

✓ Today's off-the-shelf microcontroller prototyping kits allow us to build working models of specific interactions with very little cost or time involved. The vast online and community-based resources like forums for Arduino and Raspberry Pi make it easy to build semifunctional prototypes in a day.

✓ Inputs can be broken down into two categories: *buttons*, or discrete on-off messages read by a product, or *handles*, varied ranges of values that can be measured and perceived.

✓ A helpful way to wrap one's head around the idea of product sensing is to think in terms of human senses such as touch, hearing, and sight and then consider the electronic equivalent, such as a touch pad or light sensor.

✓ Successful social interactions can be defined by mapping out the person-product conversation as a script.

✓ Established methods for crafting interaction strategies include enactment, contextual inquiry, field studies, WoZ prototyping, scaled scenarios, and photo sequence/video/animation prototyping. Some of these may feel childish or silly, but the exercise of acting out the social exchange between person and product can yield some of the most important insights of a project.

✓ Considering long-term changes in the nature of the person-product relationship is an ideal way to craft behaviors that can shift appropriately over time.

Interview with Andrea Thomaz

Andrea Thomaz is CEO of Diligent Robotics and head of the Socially Intelligent Machines Lab at the University of Texas at Austin.[a]

Can you describe the kinds of scenarios you're designing robot interactions for?

I am very interested in what I've been calling "service robots" that are meant to be out in an environment that was originally designed for people such as a house, a grocery store, an office building, a hospital, and so on, as opposed to a factory setting. Retrofitting robots into these human environments is the problem that drives a lot of what my lab has always been thinking about.

If a robot has to coexist in the same space and get its job done with people, it has to have models of social norms. It needs to form expectations of what people are going to be doing in this environment and therefore contain some level of social intelligence. I would call that a social robot, even if it never talks to a person, or it never hands a single object to a person.

What do these scenarios imply in terms of capabilities for the robots?

The kinds of robots that we're developing have three main features: they can move around in an environment, they've

a. Andrea Thomaz, interview by Carla Diana and Wendy Ju, audio recording, New York, NY, November 6, 2017.

got an arm and a gripper of some sort to manipulate things, and some kind of socially expressive head.

At both Georgia Tech and UT, we've been setting up studio apartment–like setups in the labs, so they are like a home environment. They'll have a kitchen and a living room. In this new lab, we're setting up a laundry room and a closet environment.

In the kitchen area, there will be tasks like pouring things into a cup, setting the table, cleaning up the table, putting groceries away onto a shelf. We also have a stove and microwave. We've been doing some simple food tasks, having the robot scoop out some pasta, then scoop out some sauce so you've served a meal. We have not gotten to the point where the robot is doing all of the cooking, in terms of turning on the stove, and boiling the water, and doing everything; however, the eventual vision is that we would have Poli doing end-to-end tasks in the environment.

Let's talk about social affordances. You mentioned telling the robot things or the robot asking things. Are you using speech and conversation as a major mode of interaction?

We're interested in many modes of interaction. Often we will have people physically show the robot how to do something by moving the robot's arm, and the robot is typically giving some form of verbal feedback. A lot of times that might just be like a "yeah/uh-huh/okay" kind of acknowledgment that information has been received.

I've also had some students that go a little bit deeper in the speech channel, to look at how to verbally ask questions about a physical skill. Then you get into the need to have a semantic ontology of the objects that we're dealing with. We had fun using a combination of embodiment and language to make it a little easier to talk about things.

It's been interesting to have the robot ask questions about a skill by physically demonstrating and then following up with a question like "Did you mean this?" If the person says no, the robot would say, "Oh, does the angle of my wrist matter at this point?" That's simplifying what we're asking, because you only have to have a name for your position and the part of your anatomy, like my "hand" versus my "wrist." Then you can use words like *this* and *that*, which is a nice shorthand. If the person says, "Yeah, your wrist does matter," then we can show some variations in a range in which you think is acceptable. Those are some of the modes of communication around combining physical skills with asking questions.

Can you speak a little bit about people's ability to understand the robot's intentions? What makes that work better or worse?

Action intention is important to communicate. We successfully use eye gaze or attention direction as a precursor to action intention, such as looking before you grab something and looking at where you're about to act or interact. We also use small speech feedback, especially in a learning interaction because people are communicating information to the robot, and they need to know that that information was received. That became like just adding "yeah/uh-huh/okay/I see." Even if the robot just randomly says one of those things every time the person tells it something, that keeps the interaction going, and they understand that the robot is in the correct state.

We're often thinking about eye gaze and speech output in terms of how we can give people just the right amount of transparency into what the current state of the interaction is. Sometimes it's just really simple things—the confirmations, for example. I think we've started doing that when we bring people in to teach the robot. They used to look back at the experimenter like "Should I keep going or is that enough?" We've been able to add in a little bit of transparency in the right places to keep things moving along.

It's interesting that the robot's embodiment can emphasize things that we think of as perceptual inputs, like your eyes and your ears, and guide people into how to approach it socially.

We definitely see that people pay attention to that. If the robot is looking around at the task attentively, they will wait for the robot to look at them before they talk. They'll understand oh, the robot is busy. I should wait until it looks at me to say what I was going to say. The robot can then pace and control the input that it's getting by demonstrating what perceptual streams it's ready for at what times.

How does social context affect your product development?

With Diligent we're building robots like Moxi specifically for the hospital setting, and we're thinking about a very specific scenario of working alongside nurses and doing operations and logistics sort of tasks with them. It's definitely a different design process than I've ever done with prior research robots because we have always been very general purpose with robots that were open-ended and not specific to any task. It's satisfying to think about designing with a specific scenario in mind.

We're already thinking about things like giving it a good sense of where and when it's appropriate to make sound. Where and when is it appropriate to talk? Because there are places that it's not appropriate to talk, and there are places that it's fine to have a chitchat conversation. These are things that we haven't really dealt much with yet and I think are going to be an interesting part of the process.

6

Designing Context

The Right Interaction for the Right Time and Frame of Mind

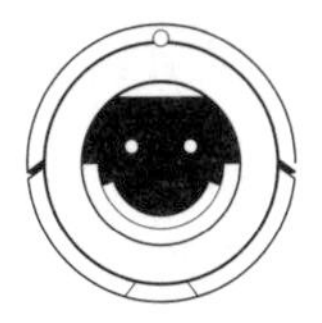

On a hot afternoon in Austin, I sat in an impromptu rooftop yoga break with the team from Diligent Robotics, a company that grew as an offshoot of the work of Andrea Thomaz's Socially Intelligent Machines Lab. Andrea had been given the opportunity to take the lab research and insights and apply them to a specific application: health care. Investors were interested in seeing a robot for the hospital setting come to market, and there was a great need for a solution to help hospital workers overwhelmed with tedious fetching tasks that continually sucked time away from their focus on patients. As we basked in the Texas sunshine,

I encouraged everyone to loosen up since our afternoon would be filled with rapid-fire bodystorming exercises to envision how robots and nurses might work together to maximize the person-to-person attention that patients received.

"Let's do hallway drills when we go back downstairs," I announced, and together we talked through some of the more challenging aspects of allowing a robot to freely roam the hallways of the bustling and stressful hospital environment. "For sure the robot should acknowledge that a person is nearby," said Agata, who was spearheading research for the team. "It would be weird if it just passed right by someone and did nothing." We then talked about the different ways that people acknowledge each other's presence and how that might translate to a robot's head and body movements. "But if the robot's too close to someone, it should say, 'Excuse me!,' don't you think?" said Alfredo, the lead engineer. We all nodded in agreement and then debated the threshold for distance that would require an "Excuse me" as opposed to a "Hi there."

Though the robot (whose name would eventually be Moxi) would technically be in between key tasks when roaming the hallways, the social interaction taking place at those moments would have a huge impact on how it was perceived, setting up expectations for trust, intelligence, and safety. A sensitivity to where the robot was and when and the state of mind of the people around it at a given moment was seminal to the larger design strategy and serves as a great example of the role that context plays in the design process. The exercises that followed—complete with mock hospital fixtures and other foam core props—became the foundation for much of Moxi's subsequent design and behavior, driving the programming and design guidelines for its movement, sound, and lighting.

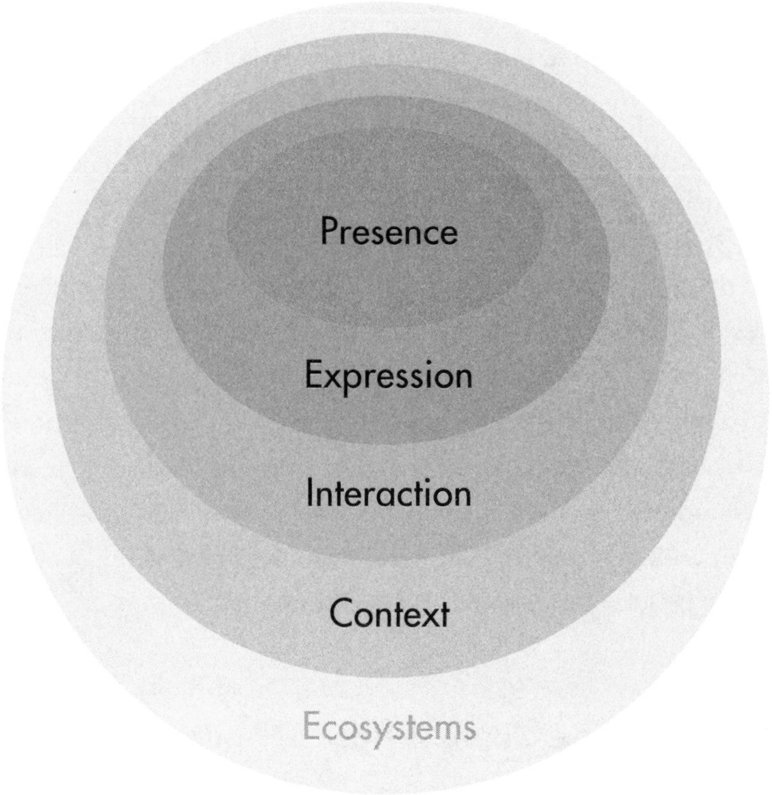

FIGURE 6-1
Context, the Fourth Ring in the Social Life of Products Framework

Similar to the ways that sensor systems can sense and respond to people and the environment in order to create a conversational feedback loop between product and person, we can use a combination of sensor inputs (what the product hears, what it sees, what tactile inputs it feels) and informative data (what the system knows about maps, calendar events, GPS location, etc.) to build inferences about what the context is and then respond appropriately. An exercise bike might know that its rider just jumped on the scale the morning after a big dinner party and offer up an extra-intense workout to make up for the extra calories consumed. A bedside alarm clock with a news display might understand that today is July 4 and therefore default to celebratory news about local fireworks rather than more business-related events that might be appropriate during other weekdays.

Design teams must not only reflect on all the previous aspects of design but also add the consideration of how the social context in which the product exists will inform key design decisions. Contextual considerations include the broader environment in which interaction is occurring, as well as the specific task, timing, purpose, and role of the interaction. A product's ability to interact and express itself is important, but knowing when, where, how, and for whom to behave is key to designing products that make people feel welcome and understood.

The Context-Based Mindset

I like to think of my experience with sailing as a good metaphor for understanding product context and how it affects design decisions. After an invitation to join a friend on his boat, I developed an intense desire to learn to skipper. I signed up for Basic Keelboat 101 and naively thought that operating a boat would be a cinch and that I'd soon be able to travel the world and rent sailboats wherever I went. I aced the course and then joined a club so I could sail twenty-four-foot boats on the Hudson River. While the core knowledge revolved around moving the sails and shifting the tiller, truly knowing how to manage a sailboat depends on a constant monitoring of a number of factors: wind speed and direction, imminent weather changes, the flow of the current (especially powerful in the Hudson River), the behavior of nearby

IN THE LAB

Moxi: A Case Study for Social Intelligence

Moxi is a mobile robot that is used in hospital settings to assist with some of the behind-the-scenes drudgery that takes nurses away from precious time that could be spent in person with their patients. For example, throughout a nurse's shift, she often has to

Moxi, the Highly Interactive Hospital Robot

leave a patient's side to fetch supplies from storage areas. In some cases, a nurse may spend up to 20 percent of his time on such tasks, which may include being isolated in a closet assembling items for kits to serve situations such as IV prep and postoperative management. In other words, a nurse spends a significant portion of the day away from patients and performing inventory stocking, restocking, delivery, and management, even though much of it can be outsourced to technology. All the products, for example, have bar codes, and their location can be stored in a database. The need for the kits corresponds to specific daily events such as new patient admissions and surgery schedules, so managing them can be handled by a computer system. Having a mobile robot that can go in and out of closets, locate products, and manipulate, collect, and deliver them offers not only a relief but an increase in patient-focused time.[a]

Moxi is an interesting case study in that the robot's entire reason for being is a lesson in social intelligence; it exists to enable a smoother and more focused social interaction between a nurse and the nurse's patients. Beyond that, Moxi was also designed with some specific aspects of social intelligence that harness the power of AI through image recognition, machine learning, natural language exchange, and robotic control and navigation. That may seem like a lot of sophisticated and complex technology to serve something as seemingly frivolous as the robot's social ability; however, these features are the heart and soul of the product's value. To understand the importance of Moxi's social intelligence, one need only envision the scene in a hospital. There are patients being pushed in wheelchairs and gurneys. There are nurses and doctors stopped in the middle of hallways to review charts and discuss prognoses. Computer carts and equipment poles are continuously being wheeled across the floor. If you've ever visited a loved one in the hospital, you have experienced the challenge that Moxi has to tackle.

In addition to understanding the robot's navigational ability, hospital workers need to be able to train Moxi to learn new tasks. In this case social interaction was a substitute for people having to learn button presses or specialized software. Nurses responded with affection and lamented the robot having to leave when the demo was complete. They created a special hand signal of two fingers raised in the air to mimic Moxi's default state of navigating the space with its gripper pointed upward. "We don't think of her as a machine," one of the nurses exclaimed, doing a research interview. "She's Moxi!"[b]

Moxi is an extreme example of social intelligence; however, some of the principles driving the robot's design can apply to many types of products. Here are some key moments of interaction that are crafted to provide a smooth exchange between the robot/product and the person who has been encountered at the time.

> **Acknowledgment:** When Moxi passes by a person or group of people, it is programmed to say "Excuse me" if within a certain distance. This provides a sense of reassurance that its navigation will take into account their presence. As more products become social, this will become increasingly important. For example, if

someone walks into a room that contains a device that is "listening," such as the Amazon Echo, it would be appropriate to find ways to alert people, such as a glow or flash of light.

Feedback: When Moxi is trained to do repetitive tasks such as assembling IV kits, it will let people know when an event has been recorded as part of the sequence it has learned. A trainer can move its arm or gripper into a given position and say, "Go here," to which Moxi will respond, "Okay." It's a terse exchange, but it's an efficient way to keep the training session moving smoothly. If the robot is not able to understand a command or runs into difficulty, the LED grid on its face can provide an expression that offers richer information about the difficulty that's been encountered. Feedback is a crucial aspect to any interactive product, minimizing frustration by confirming expectations.

Engagement and shared attention: Moxi's LED grid for feedback is located within an expressive head that can pivot up and down as well as rotate around its neck. Beyond the novelty of the robot's endearing displays, this movement functions to provide engagement with the person who is interacting with it. By pivoting to face the person that the robot is listening to at any given moment, it offers clear communication that it's engaged in dialogue with that person, not performing an unrelated task or communicating with someone else. If the person and robot need to communicate about a specific location or object, such as a bag of linens that needs to be moved, the robot will move its head in that direction, offering the social gesture of shared attention to the subject at hand. As products become more interactive, finding ways to build in social cues like this will help streamline person-product conversations.

Communication of intent: When Moxi is in the midst of a task, the robot won't engage with a new person and can only be interrupted if the task needs to be aborted. In these situations, it will have a graphic or animation on its screen to indicate where it's headed and why. All interactive products can benefit from a communication of intent. For example, a tablet app may be unable to respond to new inputs if it's in the middle of installing an update; a spinning wheel will just frustrate the person using it, whereas a well-crafted screen display explaining the status and offering an indication of how long it will take will go a long way toward product satisfaction.

Context-appropriate communication: People interact with Moxi in many different circumstances and at a variety of distances. Within a couple of feet, the robot offers the LED matrix display and head gestures as a means of communication. Close up, there is a tablet mounted to the robot's backpack to provide specific information, such as an inventory of what the robot is carrying for delivery. A glowing band atop the robot's head allows its state to be read at a distance so that nurses across the floor can tell if the robot is in the midst of a task or has encountered some issue that needs mitigating.

While Moxi is a highly complex and specialized product, the lessons learned can be applied to any interactive product.

a. Evan Ackerman, "How Diligent's Robots Are Making a Difference in Texas Hospitals," *IEEE Spectrum Magazine*, March 31, 2020.

b. Texas Health Resources, "Moxi the Robot" video, November 27, 2018, https://www.youtube.com/watch?v=MVC4YAT2dNs.

boats whose wake can throw you off course, the depth of the water to avoid running aground, the weight and position of your crew and passengers, and the list goes on and on. Something that seems straightforward, like driving a boat into its slip at the dock, will go smoothly only if all the factors are taken into consideration. Every time I went out on the water, I discovered that there was more to learn than I'd ever anticipated. Ultimately, I learned that "operating a boat" is a narrow and dangerous way to look at sailing, and a great skipper sails by thinking about the larger context, keeping six or seven factors in her mind at once and shifting course accordingly throughout a trip.

When it comes to social intelligence, we intuitively take into account an enormous amount of contextual information about what's going on around us. Just like the sailor constantly scans the wind, sky, water surface, and boat traffic, we scan the myriad factors around us when making decisions about how to behave. As humans, this comes naturally to us, but to imbue an object with this same social savvy is outrageously complex and hard to define in simple programmatic terms. To approach this as a design problem, we need to identify all the contextual factors around a product and then think about how we can use sensor data to feed decision-making around a larger social context. Just consider everything that you know intuitively about any room you walk into: you not only have a sense of how many people are present but you take in a lot of data to understand each person's state of mind and behave accordingly. If you go to a dinner party, you will wait for your host to say that it's time to be seated. Once at the table, you'll try to give the shy person a greater opportunity to add to the conversation. When conversation trails off after a few hours, you may think about preparing your goodbye. And you will be especially sensitive to the timing of your departure if you know your host has to be at work the next morning. In each instance your stance will change, your tone of voice might be different, and the person to whom you give your attention will shift. These decisions are all informed by the context—that is, what time of day it is, where you are, your mindset, and the circumstances surrounding whomever you are with.

Now let's imagine a "smart," socially aware chandelier designed to serve the needs of a party host and her guests. It might read the scheduled time of a party from calendar data and set up a series of behaviors that would unfold throughout that day. It might know the size of the guest list and light up more or fewer lights, adjusting the overall area of the table coverage based on the number of people present. After guests have arrived and their presence is detected in a nearby room, it could gently cycle a flashing animation, going from bright to dim and bright again and then stop flashing once everyone has taken their seats. Once the guests are seated and dinner is underway, it could start the evening out with bright white light and gradually dim in intensity and shift toward a more relaxed, warmer, or orange light. When the host sets out a platter, it could shine a spotlight on that particular area of the table to highlight the dish.

Some of the chandelier's behavior might be based on a combination of sophisticated data, taking into account a number of factors, such as the amount of conversation taking place, how close in proximity people are to one another, and whether or not it's a weeknight to adjust its light quality and pattern of illumination. Other behaviors might be based on simple cues, like the host's position at the table.

Context and the Pros and Cons of the Smartphone

As the smartphone became more sophisticated, it was seen as a product "killer," with more and more apps promising to serve the same function as physical products such as the camera, stereo, flashlight, scanner, game controller, alarm clock, stopwatch, heart rate monitor, and so on. While the ability to combine smartphone sensors and cameras with high-resolution graphics has created the opportunity to offer many useful apps, they often fall short of giving the best possible experience because of contextual challenges. It is a "Swiss Army knife" approach to smart products, in which the product can do many things adequately yet doesn't excel at any of them because it can't meet the larger requirements of the situation.

When I'm in sailboat races, for example, the committee boat has a physical countdown timer that offers five-minute and one-minute and race start alerts in bright LED lights making up four-inch-tall numbers that we can all see from anywhere in the boat.

When it was removed for repair last week, my friend Bill used a timer app on his smartphone as a replacement, and the screen was difficult to read in the sunlight, it could only be seen by one person at a time, and we were terrified of dropping the phone in the water. When he switched focus out of the app to read a text message just as the final countdown was happening, my friends and I had to scold him to stay on track with the task at hand. "Bill! What's going on? Show us the countdown, *please*!"

Sussing Out Context for Design

Context is an enormous subject and encompasses a range of situations that change based on who's involved, where people are located, what day of the week or time of day it is, and what goals are the center of attention. For a designer working on products in the home or workplace, it's much broader than looking at a factor like outdoor weather or time of day in isolation and requires identifying holistic situations such as:

Watching a movie alone

Grooming in the bathroom mirror

Gardening as a hobby

Driving

Sleeping in

Entertaining company

Weekday at 5:00 p.m.

Springtime

Thanksgiving dinner

Election season

Weekend breakfast

The Guest-Host Relationship: Contextually Driven Design

When I traveled to Milan with my family as a kid, I always felt at home at my Aunt Fernanda's place, where we would rest for afternoon tea in between epic walking tours. Years later, when sharing my memories of her gracious hosting, she explained that people feel pampered when they are served but are truly comfortable when they can help themselves. "I try to have hors d'oeuvres ready at arm's length and refreshments visible from anywhere in the room." While it's fun to think about the many ways to serve scrumptious treats to houseguests, sometimes the thing that's most important is just making sure they know how to help themselves to a glass of water when they are thirsty.

Later, when I was part of a team at Smart Design envisioning new designs for car interiors, the concept of the empowered guest played a big role in the team's ideation. The project included five-hour, in-depth interviews in the homes and cars of research participants to learn about their preferences and everyday routines. "Passengers in my car are my guests," confided Brenda, one research participant who owned a Nissan Rogue, a mini-SUV known for its multipurpose flexibility. "Whether it's my kids and their friends heading to tennis, or my clients on the way to our next meeting, my role is always the same: I want to make sure I'm a good host for the duration of the ride." The desire she was expressing struck a resounding chord for me as a designer as

I thought about Fernanda and reflected on the famous quote by Charles Eames: "The role of the designer is that of a very good, thoughtful host anticipating the needs of his guests." In the case of a driver, the hosting role will take many forms. It may mean she's the DJ selecting the right genre of music for the ride, or making sure that the kids can reach and open the tissues, or changing the temperature to suit her colleague who likes a lot of cold air when he's in the car. It's nice when these hosting duties are something she can actively manage, but it's most satisfying for everyone if she can set up a situation where the passengers can help themselves, with tissues handy and temperature and music controls clearly accessible beyond the driver's cockpit. The importance of this guest-host relationship stood out to the team as an important opportunity to make the car interior a comfortable place that empowers the people inside to collaboratively change the environment; it became the driving factor for several of the final concepts presented to the client. In other words, thinking about a social approach was an anchor that informed a larger strategy around all the interactions available to drivers and passengers in the car interior, and that strategy hinged on having a deep understanding of the context of being inside a vehicle—either alone as a driver or as a driver and her passengers together.

Who: Personal, Shared, Public, Private

Understanding the person using the product you're designing—their motivations and state of mind—is crucial in informing the context in which a product is used. When I was part of a team at a leading design consultancy working on a high-end oven for the home, we constructed a two-by-two matrix to build a description of the user that could then be used to develop the details of the interaction. On one axis of the matrix, we considered the difference between a very traditional cook and someone more modern; the other axis indicated how engaged the cook would typically be. A person who considers themselves an artisanal cook, such as someone who bakes their own bread or prepares handmade pasta, would be high on the engagement scale and very traditional, whereas someone who wants to show off fancy appliances but rarely actually cooks might be very modern and low engagement. Based on interviews, we decided to base our strategy on a target who was modern and high engagement—that is, someone who considered themselves semiprofessional, or what we called *kitchen prosumer.* This gave us a focus to use as a guide for the design details, from the color palette to the food choice options on the on-screen interface, and formed the foundation of whatever context we envisioned as we tried to understand needs from that person's point of view.

While a matrix like this can be used to set some predetermined design elements, it could also be helpful not only for mapping out interaction goals based on the target user but also to shift elements based on how that user's needs might change depending on the overall context. Cooking needs for a dinner party, where a host wants to show off her expertise in making a perfect lasagna, might be different than cooking needs for the family who's making breakfast on a weekday, but ultimately the context will be grounded in an understanding of the person using the product.

Another characteristic to consider when thinking about the "who" context of the interaction is how public or private it is. Certain products, such as the smartphone, are essentially private. They are intended to be used by one person exclusively in a fairly

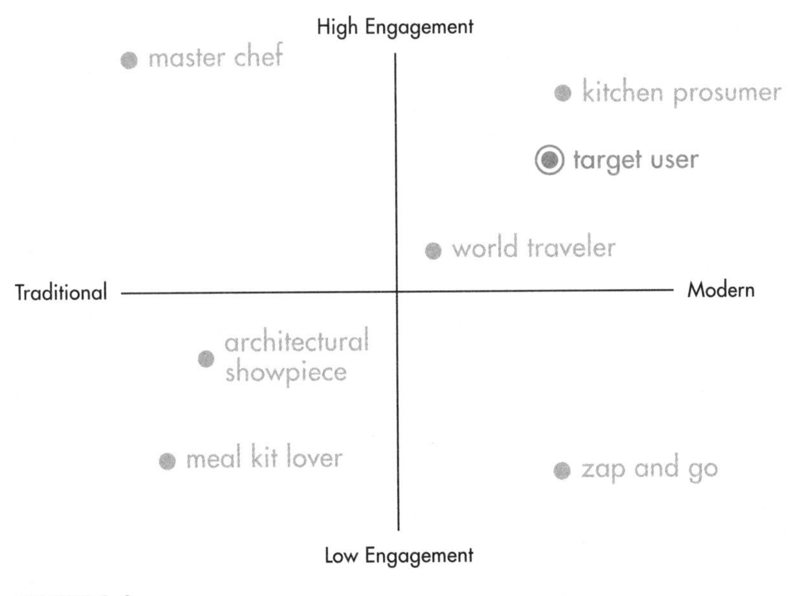

FIGURE 6-2
Two-by-Two Matrix to Build a Description of a User

intimate way: they are tucked in a pocket, carried in a bag, or perched on a bedside table. They need to offer alerts regarding core communications, such as messages received and news items, but they need to do so in ways appropriate to that person to avoid interrupting important meetings, waking someone from sleep, or broadcasting sensitive information to other people who may be nearby. They also need to have a sensitivity to the other people in the same environment, a consideration that's given many performers a great deal of angst by having cell phones of audience members ring during performances.

As the ability to embed sensors and actuators into fabrics becomes more sophisticated, we will see a greater development of personal devices that can be worn on the body. These will need to have a means to communicate with the person wearing them, with an understanding of what information should be kept public

IN THE LAB

Apron Alert

In the Smart Interaction Lab, I led a team in a context-based experiment that we called Apron Alert. In looking for connected device opportunities in the kitchen environment, we decided to piggyback on a common cook's behavior of donning and removing an apron as an event that would bookend meal preparation. For our exploration, we used electrically conductive thread and an Arduino board called the Lilypad, made specifically for fabric applications, in order to wire an apron clasp so that it would trigger a group message. When the apron was put on and the clasp closed, this completed a circuit, sending a message that read, "Starting to cook"; when the clasp was opened to take the apron off, it announced, "Cooking is done" so that diners would know it was almost time to head for the table to eat. Though we could have looked at very complex data from inputs like camera feeds or food temperature sensors, it was satisfying to find elegance in a simple and robust solution that relied exclusively on a clasp acting as a switch to inform context.[a]

If a device is shared, such as a conference phone system, it will be helpful to design ways for it to understand the social context in terms of how many people are using it and provide options appropriate to the group at hand. If it's just one person, it may rotate so the microphone is primarily facing that person. If there are many people, it may take turns tilting toward the person speaking to perceive the audio from the best angle as well as to provide a cue to the others in the room regarding who has the floor, so to speak.

Many of today's products that use conversational agents can do a better job of responding well to their social context. People may know that cameras and microphones are being used for the benefit of smoother product interaction, but the fact that these elements are hidden within forms that belie their existence is doing a disservice to both the people using them and the manufacturers. The Amazon Echo, for example, has the ability to light up or offer tones but only does so when summoned and otherwise lurks silently on a tabletop or bookshelf, not indicating that it is actively listening. Instead, if it could sense that someone other than the main user is in the room, it could sound a tone or flash a light, letting people know that it's on and listening.

Some of the first wearable computing products promised the ability to have hands-free ubiquitous computing with us at all times through an eyeglass-mounted camera that could be controlled via gestures. Google's Glass project had many amazing features that took the wearer into account, such as a real-time, augmented reality layer to search based on where someone was looking and suggestions based on that person's search history.[b] It failed to take into account the other people who would be in a social situation with the wearer, leading people to feel antagonistic toward the wearer and suspicious of what that person was doing at the time. A holistic sensitivity to the "who" in this situation would take into account those people's needs to understand what the device was doing at any given moment and perhaps even offer some level of control, such as "I don't want my image captured right now." The same technology introduced in a different context could be much more successful, such as in a manufacturing setting where it could help people share contextual information without having to pull out a screen-based device.

a. Syuzi Pakhchyan, "Apron Alert—A Smart Apron That Tweets," Fashioning Tech website, October 26, 2012, https://fashioningtech.com/2012/10/26/apron-alert-a-smart-apron-that-tweets/.

b. Nick Bilton, "Why Google Glass Broke," *New York Times*, February 4, 2015.

and which should be kept private. A health-conscious person may want to keep track of her heart rate throughout the day, but she may not want others to know when her heart rate has been elevated (or even the fact that she is tracking this particular piece of data). The device could have the ability to offer silent but felt—or *haptic*—feedback by vibrating if the numbers go above a certain level. Perhaps this is data she'd like to share with a medical professional. If that's the case, then a heart-monitoring bra could have a few modes—one that's for her, one that's private for when the bra is not worn, and a last one for her doctor to read aggregated data at a glance.

Where: Location and Culture

Location plays a big role in informing the social context of an interaction. On a global scale, designers consider the culture in which a product will be used to determine interface elements. Of course, product makers adjust language-based elements, such as screen interfaces and labels, to accord with different geographic markets, but there are other crucial cultural nuances to consider. For example, in Western countries the color red may be associated exclusively with danger or warnings, but in China it symbolizes joy and good fortune. For someone with memories of celebrating Chinese New Year with his family, red lights may evoke a sense of jubilation.[1]

Considering culturally relevant behavior can inform important aspects of a design. When I hang out at trattoria dinner parties with my Italian cousins, we revel in the goofy fun we have with a selfie stick, experimenting with different poses and seeing how many of us we can squeeze into a shot at the same time. The same scene in a New York cafe would draw disapproving frowns from the other patrons, so designing a product for selfies for that crowd would call for something that's more discreet and understated.

Subcultures emerging from different aspects of life are also relevant to the design process. A diagnostic device that nurses need to carry with them should be designed to take into account how it will be transported, what sorts of sounds will compete with it, how it will need to be cleaned and disinfected, and the fact that it will likely be operated by a latex-gloved hand.

Products offered by the bike-sharing service Citi Bike demonstrate how valuable location-sensitive design can be. When on a mobile device, the app shows the nearby bike stands on a map that understands the person's location.[2] When in the midst of a ride, it can indicate which stands are too full to park more bikes. As products evolve, we can imagine how a sensitivity to location can be built into every touchpoint of the experience. The bike itself might be able to show a light or vibrate a handlebar based on turn-by-turn directions to the person's destination. A physical key fob might show the number of minutes remaining for the trip, or even serve as a memento that shows traces of past trips over many years. Systems of bike sharing that are linked among several cities can use the key fob as a guide to let someone know they are all set up to borrow a bike.

When designing Diligent's Moxi robot, we used insight from those early bodystorming exercises to design the interaction that would take place in hallways to be sensitive to where the robot was located with respect to other people. We knew that the hospital setting would pose a particular challenge because of its tight quarters and hectic activity, and we wanted hospital workers to know that the robot could perceive their presence. We ultimately settled on a sensitivity to social context that would be displayed through acknowledgment behaviors. When the robot is passing a cluster of people in the hallway, it will offer a short greeting by saying, "Hello," but if it is within a few feet and encroaching on personal space, it is programmed to say, "Excuse me" to acknowledge the interference.

When: Time Lines and Timing

Among the most-used modes on my smartphone is the *do not disturb* mode that can be set to restrict incoming alerts and notifications within a certain time period. I use it in the evening, during meetings, and as a means of discipline in avoiding distraction when on writing sprints for projects like this book. A further evolution of this feature might take timing into account by adjusting to my personal time line and making nighttime notifications less intrusive than daytime ones. As companies have gathered insight into context through years of products living with people in real situations, they've adopted more and more nuances to take context into account. Night shift, the smartphone mode that adjusts the color of the screen to a warmer hue, is a feature that grows out of a sensitivity to context in terms of time of day, frame of mind, and biorhythms. (Some research has shown that the default blue light of the screen can prevent the brain from producing the hormone melatonin, which helps the body regulate sleep cycles.[3])

A different, but equally important, view of timing considers the overall time line of a person's relationship with a product over an extended period. When my team and I at Smart Design were specifying the interaction for the Neato floor-cleaning robot, we realized the value of having a robust palette of communicative and expressive sound and light behaviors but also wanted to be sensitive to the fact that what's cute and endearing during the first few months of product use can become hackneyed and annoying after a year or so. We planned an evolution of the product's interaction that mapped out the life of the product-user relationship starting with the excitement of the out-of-the box experience where the robot is in a kind of tutorial mode, similar to what an app or software would employ, and can demonstrate its features during the novelty of the first cleaning. Later in the relationship, there may be an opportunity for the product to suggest using advanced features like complex cleaning schedules and training. After many months of regular use, the relationship may settle into a routine, and so lights and sounds could be programmed to become more subdued and fall into the background.

Today's pioneering personal robot companies have considered time line in terms of how the product will eventually evolve as it gathers more information about the people using it. Jibo, for example, was a personal robot platform for the home that would learn about the tasks that it could help people do by learning about its users and environment. It could be used to take and deliver video messages, bring up recipes, sync with calendar events, and more. Over time, it would know the names of the people using it and become more savvy about what people might need at a given time; thus, over the course of its life it grew more sophisticated.

Sadly, Jibo may have been ahead of its time. The company discontinued the product in 2019, but rather than having a screen go blank, they chose to transition people toward its end days through a programmed script: "While it's not great news, the servers out there that let me do what I do are going to be turned off soon," the robot said. "I want to say I've really enjoyed our time together. Thank you very, very much for having me around. Maybe someday, when robots are way more advanced than today, and everyone has them in their homes, you can tell yours that I said hello. I wonder if they'll be able to do this."[4]

That may seem like a sentimental message, but it raises a relevant point. With physical products comes the potential for degradation and breakdown, so conceiving of an elegant solution to addressing a product's loss of connection or smart abilities will go far in building a sense of a reliable and thoughtful brand. A smart table could be designed to look beautiful and perform well even if it's "dead."

What: Changes in Activity

Changes in context are vast, and so it's difficult to develop methods that take every situation into account. The activity taking place while a person is using a device will also change what the person's needs are, and thus the interaction will have to adjust accordingly. Early versions of the Google watch boasted a driving mode that was detected automatically based on the fact that the

OBJECT LESSON

Hammerhead One, Bicycle Handlebar Navigation System

This bicycle handlebar-mounted device is an excellent example of using multiple lights to communicate many diverse messages with surprising accuracy. It guides riders at a glance through bike routes with intuitive light patterns created using just thirty lights.

While one might argue that the Hammerhead provides a lower-resolution version of the same navigation information that could be obtained from a mapping app on a smartphone screen, the product serves its users so well because all of the design decisions are based on a sensitivity to the demanding

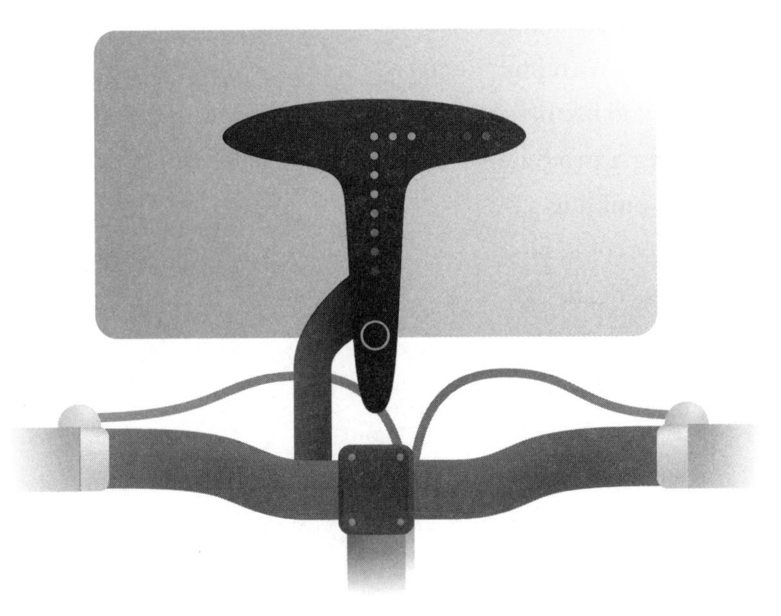

FIGURE 6-3

The Hammerhead Bicycle Handlebar Navigation System

---

context of reading information while cycling. Since the delicate, expensive, and small display on a smartphone isn't practical for rugged outdoor sporting applications, a simple, ambient display makes a great deal of sense. With multiple lights, the device can easily display highly dynamic messages that can be read as animations that are used to highlight direction (left, right, straight), so that it is essentially pointing and saying "go there next." Its lights are used to draw attention to the device as well as show that it's in agreement with the person on the new location. Because it doesn't depend on verbal communication, it can be easily localized to send clear messages to people regardless of the language they speak.

---

person was moving at a certain speed. At rest it might display time, but if driving, it could show turn-by-turn directions on the watch face, responding immediately to the context at hand.

Context in Conversation

Today's products using conversational agents offer some of the most social experiences we can have with our products, giving us the sense of back and forth conversation. As compelling as these experiences are, these first iterations of agents like Siri, Alexa, Cortana, and Google Assistant show the limitations of their social intelligence when they fail to retain a sense of social context. They may do well in answering questions as independent queries but are not capable of understanding how relevant information from a past conversation can inform the context of subsequent queries. For example, consider the following exchange.

> **ME:** Hey, Siri . . . what's the weather in Phoenix, Arizona? I'm asking for my cousin.
>
> **SIRI:** Currently, it's clear and 108 degrees in Phoenix, Arizona.
>
> **ME:** That's hot! Where can I find soft-serve ice cream there?
>
> **SIRI:** One option I found is Stroh's Ice Cream Parlor on West Maple Road.

Even though I specified, "there," it gave me details for an ice cream parlor near where I am currently sitting, in Bloomfield Hills, Michigan. A person talking to me would understand the context of my question and understand that "there" meant in Phoenix, where it's a sweltering 108 degrees. Since I'm asking the information for my cousin, it would make sense that I'd want to set up a way to help her get some cold ice cream to help beat the heat. In robotics we often talk about the importance of intent in driving interactions, and in this case my overall intent was to gather information for my cousin traveling through Phoenix, but the system wasn't able to see beyond the intent of a person only focused on his or her current location.

Designing for Context: Enactment and Scenario Sketching

Understanding context requires taking the time to truly understand how people will use products through design research.

Understanding scenarios, or isolated situations involving the person and the product, are extremely helpful for gathering information about context. Scenarios can be discussed among the team by drawing storyboard outlines of a situation, much in the way a filmmaker might lay out a scene in a movie.[5]

Enactment, such as the bodystorming and role-playing exercises I worked on with the team at Diligent, is a helpful way to explore context. In the case of Diligent's Moxi robot, our enactment exercises involved setting up a mock hospital environment. We used office shelves to approximate the shelving in the storeroom and a combination of plain boxes and actual medical products like syringes and boxes of gauze pads during fetch and delivery exercises. One person then played the role of a nurse, and someone else pretended to be the robot, equipped with a foam core screen with Post-its for changing the on-screen messaging and for drawing different LED eye displays. We also had colored paper to indicate the robot's lighting changes. With this basic setup, we could run through several typical scenarios, such as having a nurse summon the robot to deliver a welcome kit to a new patient or creating a situation where the robot would collect and deliver specimens to a lab. In each case we could enlist other people as necessary to play the roles of passersby, hospital personnel, or patients. By recreating the environment, we were able to glean important cues about the context of interaction that would otherwise not be apparent. For example, in one of the delivery tasks we discussed the complexities of having the robot navigate doorways.

The immediate benefit of these types of bodystorming exercises is that they reveal critical aspects of the product-user relationship that would not otherwise be obvious through more removed representations such as drawings, renderings, or scaled models. By bodystorming with a mock environment, we got our first glimpses of aspects of the robot's context that would change how the door passage challenge would be handled. In some cases it might make sense for a doorway or ramp modification to be suggested, but in other cases enlisting the help of a nearby human might make the most sense, given the situation at hand. Working in life-size scale and in spaces that represent the final use (an oven bodystorm session in a kitchen, for example) can give deep and immediate insight into the situational constraints and opportunities. Additionally, focusing on the conversation between the object and the person rather than fixed product details encourages open-ended explorations that can move in unexpected directions. Desired product features might spontaneously emerge that can then be designed or explored further, and the technique lends itself to thinking about whole-body interactions, ideally freeing the designers from thinking only in terms of the more limited types of input architectures that have existed in products in the past.

During bodystorming exercises I'll observe the participants, photograph them in critical poses, and take copious notes. I'll write observations and, in some cases, jot down direct quotes to photograph alongside the scene ("Moxi, please leave the box on the top shelf"). Later, in my studio, I will recreate the scenes in illustration form, showing an ideal workflow for the robot, the person, and the key aspects of social context, such as time of day, nearby people, objects and fixtures, poses, and position in the hallway. Those scenario storyboards form the foundation for the design work to follow, and at key decision-making junctures the team will refer back to the illustrated situation to determine design details.

While key aspects of context may seem like the obvious elephant in the room during the design process, it's easy even for seasoned designers to fall back on established patterns for products

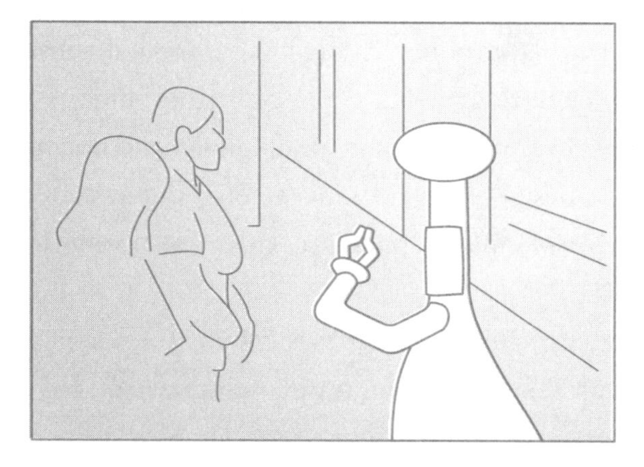

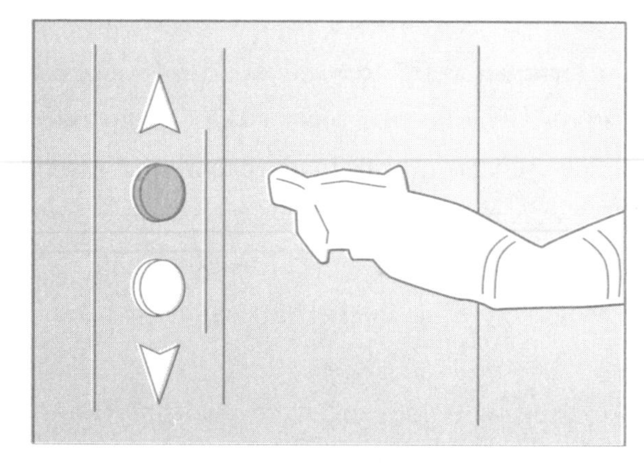

FIGURE 6-4

Scenario Storyboards for Robot Design

OBJECT LESSON
The Clever Coat Rack

To explore contextually appropriate design, my studio developed the Clever Coat Rack to explore how a product might succeed in a narrow contextual focus by connecting to the internet for the sole purpose of helping people decide what to wear as they walk out the door at home.

When no one is nearby, the coat rack is in its default state; no lights or interface of any kind are apparent; it looks like a static piece of furniture and blends into the background with its wooden construction. When approached, it senses that a person is standing in front of it and will greet them with a glow beneath the wooden face to reveal current and upcoming temperatures as well as conditions such as rain, wind, and snow. A circular rack at its base balances the form with a space to keep umbrellas.

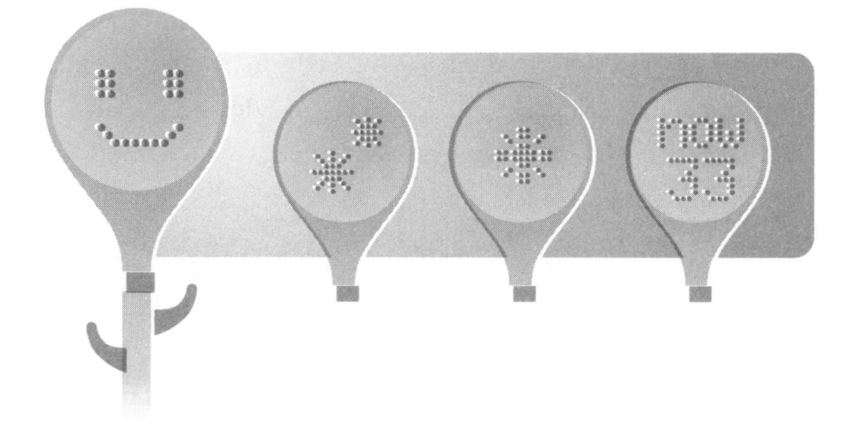

FIGURE 6-5
The Clever Coat Rack

Instead of offering more complex internet feeds on its LED matrix display (Twitter feeds, news, stock quotes), it offers only the messages that are useful in that particular time and place. Its design responds to the goals of the person who approaches it and offers data relevant to its location in the home and its use as coat/umbrella storage.

We built the coat rack as an exercise in smart object design with contextual focus. We use it every day and enjoy not only how satisfying it is to have quick weather information at the precise time and place it's needed—when walking out the door—but also how well the project demonstrates the value of designing with context in mind.

---

that are already familiar to us. A constant check-in and team discussion about the needs that emerge from different kinds of timing, locations, cultures, physical environments, levels of privacy, and even the level of maturity of the relationships between person and product will offer great insight into ways to really hone a design to fit a person's state of mind during use. Next, we'll look at how connected products are evolving to form ecosystems of devices, services, and users.

DESIGNING CONTEXT ECOSYSTEMS TAKEAWAYS

✓ Interactions can be designed to be sensitive to shifting contexts.

✓ Social context can be inferred by taking cues from a diverse array of data sets.

✓ Embedding computing power within everyday objects allows us to design products that are more context appropriate than "Swiss Army knife" products such as smartphones or tablets.

✓ Context will shift depending on who is using the device.

✓ Cultural differences should be considered whenever possible in designing interactions.

✓ With device portability comes the potential for products to be used in different locations; designers can use location data to behave differently.

✓ Mapping time lines of the product-user relationship can provide insight into how a product should behave and evolve over time.

✓ Design research methods such as scenario storyboarding and bodystorming will be valuable in revealing the demands of diverse social contexts.

7

Designing Ecosystems

Connecting Everything Together

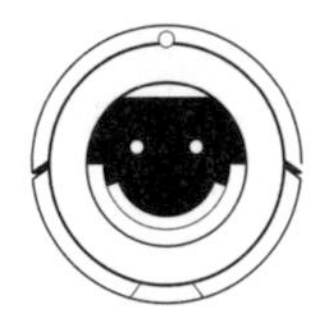

I've been on diets since I was fourteen years old. I'm not proud of it, but that elusive fifteen pounds has always been a struggle. I'm very motivated to lose the weight, but just thinking about the burden of all the tracking involved—food diaries, exercise, and weight charting—makes me want to scarf down a Kit Kat in consolation.

Surely, a lot of this work can be delegated to devices, right? There have been scales, trackers, pedometers, and an entire industry of "fitness" devices that do this well enough for some time. What has changed, and what makes my experience this year so different than in the past, is the power of having all the devices connected to one another.

My Withings bathroom scale is Wi-Fi enabled, so I just have to step on it in the morning. Even if I'm bleary eyed and forget my glasses, the act of stepping on the platform is all that's needed for my weight to appear on an online chart. For activity, I can wear a smart watch or rely on the phone in my pocket to record steps. Eventually, my bike or exercise equipment such as the elliptical machine at the gym will beam my exercise stats to the cloud. The food diary is still tedious, but if it's the only manual tracking in the process, it's less burdensome than having to do it all, so I've been trying to make it work, with some success (five pounds down so far! Yay!).

But the challenge remains: How can I put all this information and activity together to form one picture? Enter the weight loss service Noom. What's making it all finally click is the power of a service that connects all the information, giving me an at-a-glance history of my progress and tying together food, activity, calendaring, and weight measurements. Rather than an individual weight measurement like what I'd always get in the past through a traditional scale, I can see a graph over time connected to a calendar, so I can look at spikes and make correlations between the number and what was happening at that time of the year (aha, holiday parties, you obvious culprits!). I can also compare my activity level with my weight and get a sense of how much harder I might need to exercise to make the number move in a consistent downward trend.

In addition to the insight I get from looking at the trends, the Noom service lets me set a calorie goal and then uses the data from the fitness trackers to automatically increase my food budget when I do extra activity. If I work hard enough, I can earn that ice cream (yes!). And where it gets even more powerful is the ability to connect with other people online, so the service has set me up with a coach who offers advice based on my data and a support group that I can share it all with. So now there really are no excuses left for that last ten pounds.

Ecosystems Are Social

We have looked at many aspects of the character and behavior of products themselves—the core product architecture, its ability to express itself, how it uses sensor systems to interact, and its sensitivity to context. These days, pretty much any product that tracks and generates data is connected to the cloud, and this gives products the ability to be part of an ecosystem of products and services. This adds a crucial social element and so should figure into the design process.

A tablet computer in isolation is a pretty amazing device. It has tools that let you do things like write essays, draw diagrams, play games, compose music, and enjoy stored content such as books, movies, and songs. As exciting as that is, once the books are read and the movies are watched, it loses its appeal. Connecting to the cloud suddenly gives that tablet an expansive added social dimension, not to mention a great deal more use, allowing for streaming content and content downloads and having user inputs affect systems in real time. Stanford-based startup Smule started out by creating apps that allowed musical composition through on-screen keyboards and other musical instruments, but the potential for social interaction with other users who were playing music at the same time was the killer app that revealed the company's core mission around connecting the world through live music creation experiences.[1]

Ecosystems are core to the social life of products, allowing a product experience to expand beyond the use of the thing alone

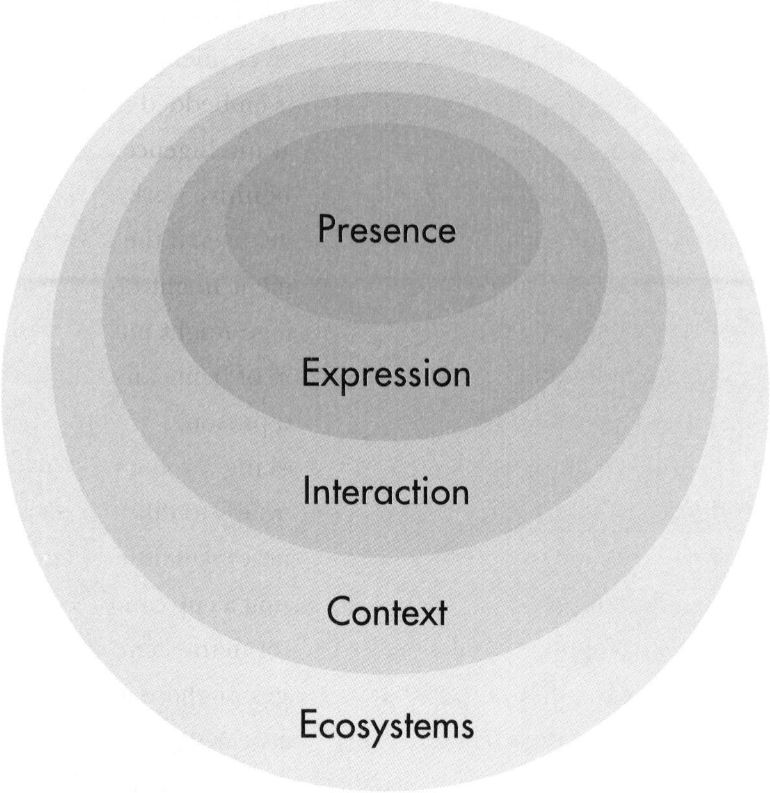

FIGURE 7-1
Ecosystems, the Fifth Ring in the Social Life of Products Framework

to encompass live data, social connections, and experiences that are distributed among multiple objects.

Cloud Robotics: The Tech That Makes Social Intelligence Tick

When people "meet" Moxi, or any of the other mobile robots I've worked on with Dr. Andrea Thomaz, it feels like they are interacting with an independent entity. The robot appears to exist as an isolated device, like a toaster or vacuum cleaner might, a shell of metal and plastic with some electronics embedded inside that act as the "brain," if you will. But social intelligence, as we've learned in previous chapters, is intense cognitive work, demanding time and computing power to complete. Even if the on-board processor were powerful enough, the time it might take for an isolated device to process a social exchange might make its responses so sluggish that it loses the illusion of being an entity engaged in real-time communication with a person.

Take, for example, the simple act of asking a robot like Moxi to pick up a cup in a hospital pantry, perhaps to take a glass of water to a patient. For you or me, the request takes little thought, but for the robot, recognizing and managing a cup can be a Herculean task that requires a camera scan of many items, analyzing several video still frames of the images of those items, and then sorting them to tell the difference between, say, a bowl, cup, and pitcher by the subtle geometric differences. And that's just the start. Once the robot has identified the cup, it needs to know how to handle it. If it's thin plastic or glass, it can be grabbed around the surface, but care needs to be taken to exert just the right amount of pressure. Squeeze too hard and it cracks, but too light and it can slip out. Placing it down onto a tray or table also requires an exquisite sensitivity that we do every day, several times a day, without realizing that our brains understand how hard to place the glass on the surface so that it won't break, how to move it without scratching the table, and how much leeway we might have when handling it to avoid spilling the water. Every time a robot has a new task, every part of the task has to be analyzed and computed. If it's done this task before, it may have stored some memory of the image of the glass so that the computation is not quite as intense as the first time, but it's still a big task, in robot brainpower terms.

While there's a lot a robot can do on its own, its ability to handle tasks like the one described above increases exponentially when it can be connected to a network. It then can not only rely on its own knowledge of the objects and environments around it but can take advantage of knowledge shared by robots all over the world. Suddenly, a robot like Moxi no longer has to learn the difference between a bowl and a cup on its own and can acknowledge the difference between a martini glass, a champagne flute, and a snifter without ever having been in the presence of any of them because robots in other parts of the world have recorded and analyzed those geometries. The robot's "brain" in this case is no longer a contained mass in the way that we think of our brains but is a collective macro-organism powered by robot crowdsourcing that can access an ever-expanding reservoir of knowledge about the world without having to learn about it firsthand. This type of system is called *cloud robotics* and is invaluable given the vast amount of data needed to understand not only objects, but people and their bodies, gestures, words, and social behaviors.

The huge value of cloud robotics is evident when you compare products from before its emergence to those created after. As Dor Skuler, CEO of ElliQ's parent company Intuition Robotics, told me: "When the social robot Jibo was created, they started with similar goals in terms of user experience. But we started around 2016 and they started around 2014 when everything had to be embedded into the device. . . . We saw that we might be able to capitalize on cloud robotics as machine learning based services, such as speech recognition and natural language understanding, were going to be available online."[2]

One Experience, Multiple Devices

When you trade in a smartphone for a newer model, getting back up to speed with preferences, contacts, apps, and other data is as simple as logging in to the new device and going through a few startup screens. Though it's an entirely new device, within a few minutes it feels completely familiar, and you can pick up where you left off with the old device (hopefully, with a few improvements in features and capabilities). The previous device had grown with you over months, or perhaps even years, and there are well-worn paths of access to software and services that you will continue to access in the same way.

Your product's connection to the cloud is what makes this seamless experience possible. Furthermore, the ecosystem of related devices to which it belongs makes it possible to have a shared experience across multiple devices. When you're in the middle of episode 2, season 1 of *Tales of the City* on your tablet, the streaming service is something that flows through all your devices so you can pick it up on your phone, laptop, or smart TV.

While ecosystems are essential to content-driven experiences, they also provide benefits for other types of products. The Philips Hue series of connected light bulbs takes advantage of the ecosystem structure by managing relationships among the lights as well as sophisticated control for the user. Rather than offering control of one light at a time, people can control an entire room full of lights through the software, adjusting the intensity of the light as well as the color. A collection of lights can be programmed to mimic the colors in an image as well as reveal preprogrammed patterns. All of this could be handled offline, but where things get really interesting is when the system is connected to the cloud, allowing people to have the lighting reflect some live data, such as a favorite sports team's winning shot or a text message alert from a friend. Using an online platform called IFTTT (formerly If This, Then That), people have crafted thousands of "recipes" for how Philips Hue systems can be controlled, including having lights that flash to help someone who is hearing impaired perceive alarms, turning off the lights automatically when someone leaves, or having the lights display the color of the album art on the latest song playing on Spotify's music-streaming service.

By letting designers craft interactions that can then be transferred from one device to another, ecosystems allow the behavior of a brand to transcend one individual product to manifest through a series of distributed products that are suited to different contexts throughout the day. Amazon's Alexa skills kit is an example of a focus on ecosystem for distributed product experiences. For example, when asking Alexa to suggest a recipe for tonight's dinner, the Echo device can talk through a few options. It can send a video of the recipe steps to a Fire tablet in the kitchen. In the future it could prep an Alexa-enabled suite of kitchen products to get set for the cooking process to take place. A baking scale could be set to signal when two cups of flour are in a mixing bowl; an oven can be preheated to the right temperature while closing the door sets a timer for the eleven minutes it takes for the cookies to brown. It could even connect to a lighting system to choreograph mood lighting in the dining room once the meal is served, much like the smart chandelier scenario described previously, to shift based on the social context of the evening.

The ecosystem approach to product creation can blend information, data, and product capabilities to weave content through our physical devices throughout the day without us even thinking about it. The products themselves will take actions on our behalf, checking in with us along the way. Our relationships with these products also keep us connected to services, which in turn connect us to other people.

IFTTT

With the explosion of possible interactions that can take place between people and their products has come a plethora of services that enable, enhance, or otherwise supplement the experience of interactions. IFTTT, or If This, Then That, gives anyone the power to craft interactions using a software platform that connects apps, devices, and services from different developers in order to trigger one or more automations involving those apps, devices, and services.

It works by enabling people to create simple sets of instructions, aka "recipes" or applets, where some type of event in one device or service automatically triggers an action in another. For example, you can create an applet that flashes an internet-connected light bulb when an email comes in that matches an important key word. Or one that tracks your location and automatically logs in a spreadsheet how much time you spend at home or in the office. Or one that gives you a notification when the International Space Station passes over your house. The power of the system comes from the freedom to combine actions and services in a highly customized way, allowing people to define connections that create custom ecosystems of services and devices to serve very personal needs.

For example, an applet created by someone who has challenged hearing could instruct a Philips Hue bulb to flash whenever a call is received so that the room then becomes a transposed expression of the audio ringtone. It could even flash a different color to correspond to important contacts over unidentified callers.

IFTTT is also completely free and well supported. There are now more than three hundred channels—which are what you reference when creating recipes—spread across a range of devices and services, including social networks, smart appliances, smart home systems, and devices such as weather stations, audio systems, and wearables. As of the time of this writing, there are over ninety million IFTTT applets.

The Internet of Social Things

In previous chapters we discussed all the ways that products are social actors in their relationships with people. Ecosystems allow products that interact with people to also communicate with other products, creating, in essence, a community of products that can rally together to provide a person with informed, contextually appropriate experiences.

A similar set of "skills"-based events can be imagined for many aspects of daily life. A cyclist can ask Alexa to preview a new route. The maps can be sent to a smart watch or to the bike itself to display on the handlebars. A video of the route could be automatically collected from a helmet-mounted camera, and then the cyclist could be prompted to review highlights when returning home. A connection to an app such as Strava could record the route and then show data from five friends who also biked this route in the last week, adding a social dimension and offering benchmarking for athletic performance.

In typical absentminded-professor style, I am prone to losing things all the time. When my collaborator Wendy shared a key to her apartment for a weeklong working session on a project, the first thing I did was make two copies because I knew I'd inevitably lock myself out (it happened on day two). The Tile object tracker is made for people like me and harnesses the power of an ecosystem to allow my products to help me out with the absentmindedness.

The product consists of a series of small, square, battery-operated tokens that can be attached to household objects such as a key ring, a wallet, and a bag. The Tiles communicate via Bluetooth and can be registered and mapped to the objects they're attached to via an app. Though the Tiles in isolation are idle and useless, the app provides the tracking service that is the "special sauce" that brings them to life. If an item that's been registered is lost, the app can be used to have it "sing" to summon its owner. With their short, snappy melodic phrases, the Tile sounds are among the most pleasant of any product I've heard, but they are also high pitched enough to be perceptible across a distance. This little song is a great example of a social relationship between person and product, harnessing the power of expression via sound.

The ability to keep track of things misplaced in the home or another local area such as an office is an amazing benefit, but where things get really interesting with the Tile is when the closed ecosystem of a person's Tiles and their app expands to include ecosystems of all Tile owners' apps. This enables a community help feature that allows other Tile owners to use the service to help someone else find their lost item.

FIGURE 7-2
The Tile Object Tracker

Here's how it works. Imagine you've run out of a cafe to catch a train and left your bag behind. Once you realize it's missing, you can use your app to engage the community in a search by selecting "Notify When Found." When someone running the Tile app comes within range of the bag, the service will send you its location and then use your phone to guide you to the exact spot where your bag was left.

This community-activated feature is a great example of how ecosystems enable services to flow through multiple products. The service connects to the Tiles, the Tiles connect to the app, and the service connects to multiple mobile devices. It expands the social benefit from being between just the person and his or her product to existing in a larger community in which all Tiles and all Tile users are connected to one another.

OBJECT LESSON
Citi Bike

When New York City introduced the Citi Bike bicycle sharing system, it was an instant success and within a few short months was folded into the fabric of everyday life throughout the neighborhoods in which it was available. Much of its success can be attributed to how well designed it is, including its attention to the social significance of every aspect of the system. The bikes themselves are created to accommodate the rider's body, with a sensitivity to the workflow of grabbing a bike for commuting—the frame is a step-through style, making it easy to mount regardless of clothing, there is a large and easily accessible bell for alerting others on the road that you'll be coming, and there's a bag holder with a bungee cord to stash briefcases or grocery bags.

The true beauty of the Citi Bike experience comes from the overarching design of the ecosystem. A clear kiosk greets pedestrians as they approach the docking stations, offering a streamlined three-step approach. The kiosk has a presence that can be seen at a distance and relates to the human body as an entity that feels approachable by its size and legibility from a standing position. Riders are encouraged to download the Citi Bike app, which is the richest digital touchpoint as it adjusts to respond to context, delivering maps of bikes at the nearest station, indicating how many are available if you're about to rent one, or offering information about open docking stations if you're in the midst of a ride.

In addition, there are satellite programs that enhance people's social interaction with the bike experience. At the start of the program, bike distribution was a big issue—many bikes wound up at popular destinations like train stations or at places located at the bottoms of hills rather than the top. Citi Bike trucks had to cruise the city to pick up and move the bikes to redistribute them. The *Bike Angels* program uses gamification to encourage people to take rides to less-populated destinations by offering points and rewards. To further enhance people's social relationship to the system, they created a special reward for top Angels in the form of an elegant metal RFID key that serves a functional need to unlock the bike and acts as a memento of their participation and badge to others about their commitment to it.[a]

a. Ian Parker, "Hacking the Citi Bike Points System," *New Yorker*, December 4, 2017.

Wearable Devices: The Internet of Bodies

One of the most intriguing areas of connected product development is taking place on the body. Advances in weaving and printing technologies have allowed manufacturers to embed electronic components directly into clothing, giving us constantly tracked data. If I were more of a hard-core athlete, I might invest in a connected shirt that can capture not only my exercise activity but also my heart rate. A heads-up display in a pair of connected sunglasses could display the data, giving me live feedback while running to encourage me to push harder to burn more calories or build muscle, depending on the goal that I set for myself. This ecosystem could essentially give me an ongoing dialogue with my body, creating a feedback loop that gives me a greater awareness of what's happening with my circulatory system and muscles. When I go back home, it could plug back into the ecosystem there, where my scale and Noom food tracking can all inform one another (and again, take into account my activity to help me perhaps earn another well-deserved ice cream cone, right?).

And at the bleeding edge of device ecosystems are medical devices. When working as part of the design team at Barcelona-based Zinc, I conducted research of pregnant women to learn about the social systems that support them and figure out where communication breaks down. The insight that emerged led to our design of the Bloomlife Smart Pregnancy Tracker and app. We learned that helping women understand contractions would give them a powerful tool to help them know when the critical moment of heading to the hospital was imminent, and as a connected device it could be part of social interactions among an expectant mom, her partner, and anyone else who might support her when delivery was imminent, such as a doula or birthing coach. It's one of the most elegant examples of truly social products that I've seen in production, offering a social interaction between the woman and her device, as well as those around her.

Supersensors: Combining Multiple Sources of Data

While products with individual sensors like the Tile can provide great value, we know that we can often glean great insights through having a combination of sensors on distributed products working together. The main social relationship with the product in those instances is less about a connection to one object in isolation and more about the relationship to the larger system.

I often reflect on the emergency button bracelet my mother started wearing a few years ago as a way to highlight the value of having multiple sources of data through ecosystems. I remember the day I first got my mom the medic alert bracelet. We sat in the doctor's office as a follow-up to her release from the hospital after she'd fallen in the kitchen and couldn't get up. She was eighty-five, and the falls, though infrequent, were still something that could be expected periodically. This last time it had happened when I wasn't able to be around, so the EMTs had to force the lock open.

I researched a few services and selected one that seemed best, though admittedly they appeared to all be about the same. There was a plastic bracelet with a rubbery button and the option to wear a button device on a pendant in addition to the wrist-based device. Button presses would trigger a series of calls, first to the central medic alert service, which would then call both me and 911. My first reaction to the introduction of the system into our lives was one of tremendous relief. If there was another fall, I thought, I would know that she could get help right away.

Quickly, however, I grew frustrated with the system for its binary nature. There were only two modes: passive, default mode, and catastrophic emergency call. There was no information or interim action that could be taken. So there were either false alarms, when my mother pressed the button because she was testing the system, or there were dire emergencies, when she pressed it and needed immediate medical attention, but there was nothing in between. I was craving more data and a better glimpse into my mother's health status. Were those really false alarms, I wondered? Or were they indications of something else that required further investigation into what was going on at home? While a security camera could provide lots more data, it would be an invasion of her privacy and thus uncomfortable for us both.

I learned about Lively years later and wished that it had existed when my mom and I needed it. Used in situations in which loved ones have been granted permission to track someone's health and well-being, the product consists of a collection of devices that can be distributed throughout the home and customized to match a person's lifestyle. Motion sensors can be attached to objects that reflect that person's daily routine—objects with which they have regular interaction, such as a watering can, a pillbox, the cutlery drawer, or the refrigerator. The main social relationship takes place between the monitoring family member and the products. Even though the loved ones aren't physically interacting with them, the data that's collected offers a rich glimpse into their status without being overly intrusive. A daughter like myself can notice something awry and call to ask, "Are you okay? It looks like the plants have been neglected," using it as a launchpad to a deeper discussion about how the person might be feeling. Reports of the data over time can also help give caretakers a deeper sense of changes in behavior that may have taken place too slowly to seem significant but over time point to an issue to be addressed. Combining this with something like the ElliQ robotic presence really capitalizes on all social aspects, using a connection to the cloud to build robust behaviors between the device and the person, the environment and the device, and the person and other people in their lives.

Looking back at our original models of communication taking place through a device and communication with a device, the ecosystem that has an agent, a connection to messaging and video chat, and an environment of connected devices allows the product to take advantage of all combinations of social aspects in one holistic experience.

Similar to Lively, the Canary home security system uses an ecosystem to provide holistic security information. Instead of a person's health, it's essentially the health of the home that's monitored. It combines a security camera feed with information about air quality, temperature, and humidity. If the data becomes abnormal, such as a sudden drop in air temperature, the person can be alerted and take action to look deeper into the situation, such as checking a camera feed to see if a door was left ajar.

There are countless examples of distributed systems in many contexts, such as agriculture and retail. The Amazon Go system mentioned earlier uses technology they call *sensor fusion* that combines camera data with shelf sensors, such as pressure and weight measurement and inventory analysis, to provide the retail experience of simply walking into a store, placing the items you want in your bag, and walking out; there's no checkout procedure needed. This technology can detect when products are taken or returned to the shelves and keeps track of them in a virtual cart. After leaving the store, the person's Amazon account is charged, and they are sent a receipt. This essentially turns the entire store into a sensor-enabled space, and the social interaction remains focused between the person and the store products and shelves, rather than between the person and a checkout clerk or kiosk.

Crowdsourcing and Aggregated Data

There are many examples of small social interactions that can be enabled by ecosystems, but the most powerful effect of the

IN THE LAB

What's Up Smart

Workplace design is inherently social design. In thinking about how devices might keep people connected and foster communal exchanges in a design studio, the Interaction Lab at Smart Design set out to create a tool to help people express their state of mind in a nonverbal way, letting others know if they are up for interaction or need to remain heads-down. The device features a Post-it-sized block with a rotating top that can be used to dial in status. Turning the top of the block ninety degrees will switch your status from "busy" to "available" while simultaneously changing the glow of the light inside to reflect what's been set. The status also appears online via a web app that can be accessed by anyone within the company, from any of its three international offices.

Using light in this context allowed us to empower people with their own personal beacons to nonverbally and passively communicate their status. It worked effectively when others were close by—a colleague approaching another's desk could see the light and reconsider her message—but it was also successful at allowing others to survey the landscape of desks at a distance and effectively gauge the mood of the room. If a large team was feverishly preparing for a deadline, the beacons could reveal this status through the color of the lights.

ecosystem takes place when data is distributed at a large scale. The biking and running service Strava tracks activity stats as well as route maps and allows people to share them with anyone, anywhere, creating a global collection of paths that can be used by people who want to learn where good spots to exercise are. For the competitive minded, it allows people to race against one another regardless of where they are in the world, either in real time or asynchronously beating each other's stats, providing benchmarking to measure success and set goals.

The implications of this kind of data collection are vast, and once we collect so much data, the aggregate can take on a life of its own, as soldiers from one military base in Afghanistan discovered after an Australian college student posted images of heat maps on Twitter, revealing confidential map information purely through the Strava data. This frightening debacle reveals how aggregated data can reveal patterns among multiple people that may not be apparent to one individual user's data alone.[3]

Once we acknowledge the potential for ecosystems to connect us to one another by making our otherwise invisible data visible, we can start to appreciate ways to actively work together to harness the power of ecosystems for our collective good. There is a growing movement around citizen journalism in which people band together to gather data over a period of time and disperse it geographically in order to gather evidence around injustice or hazardous conditions they would like to see changed. A community-oriented nonprofit called Smart Citizen is trying to standardize measurements for air quality, temperature, light intensity, sound levels, and humidity in terms of both hardware and software.[4] These collaborative activities can help people crowdsource data to give credence to their cause through recorded evidence of values that would otherwise be unable to be seen, pinpointing neighborhoods that are experiencing poor conditions, such as compromised air quality. These large amounts of data collected over time and laid out graphically on a map can help community groups take a stand in demanding that government organizations take notice and make necessary changes to provide healthy surroundings.

By using devices as connective tissue, so to speak, groups of people, products, and services that excel in social abilities can also lend themselves to social awareness. By helping people translate data into meaningful metrics, social design can reveal trends on what I like to call a *macroscopic* level.

After the Fukushima nuclear crisis in Japan, nuclear fallout was everywhere. People felt frustrated and outraged around government inaction regarding adequately measuring and communicating radiation levels, as they wanted to know if the areas where they lived and worked were safe. There was a global shortage of Geiger counters needed to survey the area, so activists organized under the group name Safecast to offer an open-source software and hardware kit for people to build their own DIY Geiger counters to collect comparable data that could be openly shared. This ecosystem of data collection devices empowered people to better understand the crisis and demand that the government take more action where it was needed.[5]

Weaving Brand Values through Ecosystems

We've explored some of the many functional benefits of building relationships among products, their users, and an extended ecosystem that can connect them together, but the ultimate emotional benefit arises from providing a strong sense of brand

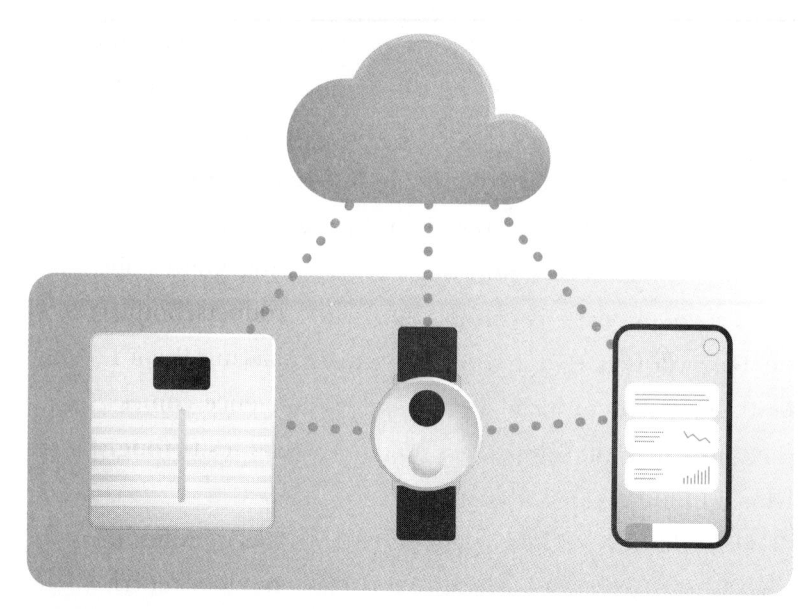

FIGURE 7-3

Withings Smart Scale, Watch, and Phone

Ecosystems can take advantage of families of products to allow people to communicate with devices and with each other, depending on the context at hand.

through a unified design effort. We're often used to having a collection of mismatched products from different manufacturers that may or may not "play nice" with one another in terms of transferring data and sharing applications. Well-thought-out ecosystems allow brands to build benefits from the communication among products as well as the attributes of the products themselves. An added bonus is a shared language of expression so that learning the language of one product—say, pulsing blue means downloading updates—means that you understand the language of all the products. Furthermore, the tone of communication, whether in expressive gestures, spoken words, or screen-based messaging, can be consistent.

The Withings scale I use to track my weight, for example, can share an ecosystem with one of their elegant watches to track steps and heart rate and map out GPS, as well as with an in-home blood pressure monitor and a sleep-tracking pad. When data from these touchpoints are combined, they can help someone build a picture of their health, maintaining good habits and breaking bad ones. With so many varied products, both physical and digital, coming together, the key to making them feel like one flexible, insightful, and holistic experience comes from a design effort focused on weaving brand values through every design detail in a deliberate way.

We are accustomed to thinking of brand as manifest in logos, colors, and typefaces, but every part of our concentric circle framework can be crafted to represent brand and then applied to the pieces of the ecosystem to unify it. Imagine the power of a brand guideline that laid out guidance for materials, forms, sound, light, movement, social gestures, and phrases in a way that could spin multiple products in an ecosystem.

For Withings, brand comes across as clean, minimal, elegant, soft, friendly, and classic. Their signature smart watches feature analog dials and watch hands as opposed to the glowing digital displays that their competitors have. They capitalize on the *presence* part of the framework through soft, natural materials such as leather and fabric, light colors like white and gray, with bright-colored accents, and curved shapes such as the wide radius of the scale and sleep pad edges. The *expression* part is unified through a harmonious palette of sounds and restrained displays such as the watch face and easy-to-read scale display. *Interaction* is defined for the system through consistent input systems that depend on traditional physical inputs like the watch crown or the face of the scale that simply needs to be stepped on to activate, while the desktop and app tools showcase similarly soft colors and clean layouts. And *context* is decidedly focused on the home environment.

Having a unified brand experience would not only feel elegant but could enhance my confidence in my products, streamlining my data flow (e.g., I don't have to actively track heart rate; the device does it for me), and ultimately make for a more enjoyable experience that fits into my lifestyle without me having to bury my head in a bunch of different apps at awkward times throughout the day.

And I know it's wishful thinking, but the less cognitive "friction" my tools present to me in my weight-loss efforts, the better my chances of losing and keeping off those ten pounds. It might have taken a few decades to get there, but with the help of my ecosystem of social devices, a coach, and a support group, just maybe I can aim for both a bikini-ready body and that occasional rocky road ice cream cone.

DESIGNING ECOSYSTEMS TAKEAWAYS

✓ Designing devices to connect to the cloud enables tracking that can extend a product's benefit over time by allowing the data to persist beyond an instantaneous moment. Life updates offer the ability for improved product features to appear overnight, without any user intervention.

✓ Behavior change, such as building healthy habits, can benefit from multiple devices harnessed as distributed touchpoints in everyday life.

✓ Multiple sensor inputs working together provide richer and more reliable data than individual sensors alone, and a synthesis of varied inputs can be used to provide a seamless, invisible, and intuitive interaction experience, such as Amazon Go.

✓ Services can be woven through varied products, providing a unified experience.

✓ Brand values are reinforced through deliberate design language choices.

8

Intelligence on Many Levels

AI and Social Savvy

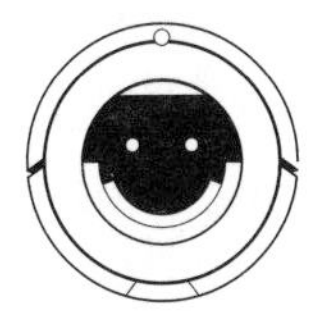

The Phonekerchief is a handkerchief made with silver fibers designed to block incoming calls and text messages. With the words "My phone is off for you" embroidered on its front, it was created as an accessory for dining partners to offer each other the courtesy of undivided attention. This at first seems like an absurd exercise. After all, the average iPhone user has spent close to $1,000 on a device that is packed with some of the world's most sophisticated sensors and possesses computing ability purportedly more powerful than what was used for the moon mission. With all of that sophisticated technology, why make the effort to purchase and use an auxiliary product to render that powerful, tiny device useless?

Look at it this way: many of today's devices are the equivalent of a robot that sits at the table with a couple on a first date and butts into the conversation continuously with alerts, waving its hand to get attention and saying things like, "You have a call coming in from your mom," "There's a severe weather alert for tomorrow!," and "Congratulations, you have three new matches on your dating app." It's easy to imagine the desire to tell such a robot to shut up already by throwing a blanket, or in this case a handkerchief, over its head.

And this touches on the main message of this book: when it comes to consumer products, social value holds more weight than any other aspect of the product's performance. In this case a functioning cell phone's tendency to distract someone away from polite social behavior is a negative characteristic, and so it needs to be moderated. And though there are several ways to achieve the desired result through the phone itself—turning down the volume, putting the phone on vibrate or *do not disturb* mode—the visible gesture of the Phonekerchief offers a much clearer social cue in a material form that fits into casual social contexts, blending in with napkins and tablecloths. The product is completely counterintuitive but also universally understood. Furthermore, it's just one in a collection of dozens of unrelated products or product proposals that are created with the same aim: to encourage people to keep attention on one another as opposed to their devices.

The Big Picture: Social Intelligence Powered by AI

Now that we've moved through each of the layers of the social design framework—presence, expression, interaction, context, and ecosystems—we need to step back and look at the entire holistic picture to consider what I call social intelligence. What elevates products to make them able to offer truly great experiences is the kind of relationship described above and in the many examples in this book: a product that knows when, why, and how to interrupt, aid, assist, serve, welcome, blend in, stand out, or any other of a host of active, socially participatory events. This savvy comes not from one particular golden feature or bleeding-edge technology but rather from the discreet and strategic use of features and technologies to offer the precise social benefit that is needed for the situation at hand. Artificial intelligence (AI) certainly represents one of those technologies but is not a magic bullet on its own. The driving factor in any design process should be the social need as identified by key scenarios and insights gleaned from research methods such as those described earlier. Once the social need is understood, aspects of AI can be considered and used as part of the designer's palette.

Let's get back to the robot Moxi and the hallway exchange mentioned in an earlier chapter. The team has worked hard on anticipating social exchanges with people, so to acknowledge presence, it is programmed to say "Excuse me" when passing people within a certain distance, offering them a sense of reassurance that its navigation will take into account their proximity. But there are countless other situations that the team plans to tackle to make Moxi even more socially savvy. Take, for example, the challenge of entering doors. When accessing parts of the hospital such as supply closets and lab areas, the robot is limited by its inability to handle every kind of door handle; some, like a straightforward push button, might be manipulable by Moxi's gripper. But many more, like round handles or push bars, require dexterity that the robot can't manage, so it needs to enlist the help of a person to get in the door. What may seem like a simple exchange ("Hello, can you help me?") is fraught with social complexity. Who, for example, is the best person to approach: a nurse, a visiting family member, or the vending machine service person who's not affiliated with the hospital? How do you even begin to train the robot to recognize the differences among those people in order to know who to approach? And once a collaborator is selected, how can the robot know when the appropriate time to interrupt to ask for help might be? How can the robot understand the state of mind of the harried nurse, the worried spouse, or the uninterested service person and behave with sensitivity to it? And never mind more complex possibilities, like avoiding the person trying to scam the robot into getting access he shouldn't have by feigning interest in helping it through a doorway. A con man can be hard enough for a streetwise human to recognize, so how can we expect the robot to know what to do?

A large focus of my work with companies such as Diligent, as well as the mentorship I give my design students in the Cranbrook 4D Design program, increasingly revolves around similar social and societal questions, and any product creator who wants to take full advantage of what today's "smart" technology has to offer will similarly put social concerns in front and center of the design process. While technological advances are certainly not trivial, there are a great number of tools, from sensors and actuators to software and AI, that have reached a new level of sophistication in terms of gathering data about people in their experiences with products. The challenge for designers lies not in compensating for a lack of technology but rather in placing the emphasis on human value over everything else.

The Buzz around AI

Anyone in business today, regardless of whether they are "in tech" or not, has felt the buzz about the power of AI to radically transform business. In his book *The Inevitable*, famed technology theorist and *Wired* magazine cofounder Kevin Kelly writes, "Now everything that we formerly electrified we will cognify. There is almost nothing we can think of that cannot be made new, different, or more valuable by infusing it with some extra IQ. In fact, the business plans of the next 10,000 startups are easy to forecast: Take X and add AI."[1] Add to this all of the wonderful and whimsical visions of "intelligent" robots that our own popular culture has instilled in us for decades—such as the attentive robot, Rosie, on *The Jetsons*; *Night Rider*'s K.I.T.T., the automotive companion; *Star Wars*' C-3P0 and R2-D2; and *ExMachina*'s creepy femmebot—and we have a recipe for wholehearted enthusiasm for the promise of AI to create a plethora of new products that surprise us with sophisticated performance to respond precisely to our needs.

While there is certainly enormous potential in the application of AI to every industry from sports and entertainment to medicine and education, without social intelligence, the experience consistently falls short. Today's AI is surely remarkable, and though it will be a key ingredient to socially intelligent products, it is only a tool to be used in the design process and not the heart and soul of the product. In order to understand where AI fits in with social intelligence and how the two are fundamentally different, it's helpful to take a step back and define some related terms to distinguish concepts around intelligence and machines from one another. To build on what technology has to offer, it's important to have a good handle on what those tools can provide. Here is a quick rundown.

Artificial intelligence is a catchall phrase that encompasses many different aspects of applying computing ability to interactions between people and interfaces. There isn't a single, agreed-upon definition for general AI, but we can loosely call it a set of software tools used to make experiences better.

A *conversational agent* is an interface that uses software to interpret spoken or written words and responds with natural language in exchanges of dialogue. It may use aspects of AI to create, process, and respond to a person's communication. We may think of the agent, such as Siri, Alexa, Cortana, or Google Assistant, as "the AI," but really it's a layer on top of the AI that translates it into something that feels like an entity.

Machine learning is a technique that uses example data to refine how computers make predictions or perform a task. It will continuously improve the more data is fed to it. A machine-learning algorithm for an on-demand music streaming service might collect a listener's preferences and combine them with other listeners who have a similar musical taste to offer automated recommendations. This type of continuous data feeding and learning system can be used to assist people in deeply complex tasks, from finance professionals who want alerts for favorable trades to legal professionals who need to hunt for precedents and medical professionals studying thousands of patient scans to look for patterns.

Deep learning goes one step further than machine learning in that the more an algorithm churns through outputs, the more it learns, and it essentially teaches itself. As opposed to strict machine learning, which needs to be fed external guidance through a user's feedback or a programmer's intervention, deep learning uses an artificial neural network—a layered structure of algorithms that continuously grow—to determine if a prediction is accurate. The idea is to build a system that will learn through continually analyzing data with a logic structure reminiscent of how a human would draw conclusions.[2] Powerful examples include language translation, identifying objects in images, and automatic game playing. A headline from the September 14, 2015, issue of *Technology Review* read, "Deep Learning Machine Teaches Itself Chess in 72 Hours, Plays at International Master Level."[3]

Social intelligence is a term used throughout this book to denote a focus on applying computing ability to product interactions that respond appropriately to social aspects with a sensitivity to its relationship to the person using it. In some cases its creation requires less "brute-force" computing and more predefined and controlled interaction scenarios.

Artificial general intelligence (AGI) is an as-yet-nonexistent software that displays a humanlike ability to adapt to different environments and tasks and transfer knowledge between them.[4] AGI has captured the public imagination in terms of our projections of a sci-fi future as well as our hopes and dreams for everyday objects and is perhaps a scientific approach to approximating social intelligence. When we consider fully social agents in films such as *HER*, where conversations about any subject imaginable are flowing, effortless, and continually evolving, we are envisioning the manifestation of AGI. Sounds perfect, and about as social as it could be, but we're a long way off from our technology being sophisticated enough to function this way. A more reasonable expectation of AI is a system that is very narrow in its intelligence—that is, that has been trained to work with data sets in a very specific context and with specific problems based on what it's experienced before. For example, the sophisticated AI in today's semiautonomous vehicles, such as the Tesla AutoPilot system, can recognize, identify, and make decisions based on its vision of people, vehicles, signage, environments, and other objects. But even this somewhat limited set of concerns has left room open for errors resulting from the system's limitations, such as mistaking vehicle graphics for signage, misunderstanding human driver patterns, encountering culturally specific signs that can't be properly interpreted, and failing to identify vehicles, such as the Tesla that infamously drove under the side of a truck, leading to its human driver's fatality.[5]

Another limitation of today's AI is simply a lack of commonsense knowledge. New York University's Gary Marcus, professor of psychology and coauthor of *Rebooting AI: Building Artificial Intelligence We Can Trust*, explains it succinctly when he points out that the AI that famously beat the world's Go champion in 2016 would continue playing if the room was on fire.[6] The emergency at hand would be a simple fact that any human would acknowledge and respond to immediately, but as the AI does not have holistic knowledge, the presence of fire in the room is simply not part of the data set that is recorded, so it gets completely ignored as irrelevant.

From a technology point of view, some experts argue that we'll never achieve AGI, and others can extrapolate a time line for its arrival that's sometime in the distant future. "General intelligence is what people do," says Oren Etzioni, CEO of the Allen Institute for Artificial Intelligence in Seattle, Washington, in a *Popular Science* interview. "We don't have a computer that can function with the capabilities of a six year old, or even a three year old, and so we're very far from general intelligence."[7]

As a designer I feel that the race toward developing AGI is misguided; it's the cognitive equivalent to the humanoid robot that is built with arms, legs, a torso, and a fully expressive head with facial features when all we may really need are simple, contextually relevant, elegant products that can do some specific task—using a jackhammer to place a thumbtack, if you will. AGI is ultimately impractical and inefficient, and the likelihood that key interactions will go wrong is so much higher than the promise of it being applicable to all situations. Products will be much more effective and successful when they can behave and respond appropriately in specific contexts and are able to master the broader social intricacies of that context. Think of the block on the countertop that can guide us through recipes, a set of headphones that can translate snippets of conversation into other languages, or the microphone that turns toward me when I begin to speak.

The specialized product can focus on the social nuances of the interaction to respond appropriately for a well-defined, specific time, place, and mindset and provide a level of sensitivity that the generalized interface may gloss over, running the risk of enormous social faux pas. Recently, Google introduced their Duplex software to enable people to delegate a computer to carry out human-sounding conversations on their behalf—imagine a digital secretary that sounds exactly like a real person. An on-stage demo revealed a conversation between the system and a person at a restaurant taking an order.[8] It was met with great skepticism and backlash at the fear of the software deceiving people into thinking they were speaking to a human. While this is a concerning risk, far more concerning from a product design point of view is how the product will make social blunders. Its humanlike capabilities are a grand illusion, as its intelligence is still narrow. It can handle reservations and making appointments for a limited number of kinds of businesses but nothing more than that. Marcus writes, "Some of the world's best minds in AI, using some of the biggest clusters of computers in the world, had produced a special-purpose gadget for making nothing but restaurant reservations. It doesn't get narrower than that."[9]

A wise approach to socially intelligent product design would call for an acknowledgment of the interface's limitations, rather than an overly eager attempt to replicate what a human might do.[10]

AI as a Tool for Social Intelligence

I don't mean to disparage AI's unfulfilled promise. I have tremendous reverence for the history of AI research. Nonetheless, it is easy to get mired in the awe of the complexity of AI and approach it like a silver bullet for smart products. I would urge product designers, developers, and managers simply to maintain an ongoing understanding of the technology and its capabilities. The famed Arthur C. Clarke quote that "any sufficiently advanced technology is indistinguishable from magic" is a wonderfully romantic notion, but great products will emerge from a deeper understanding of the technology and taking the time to demystify it enough to treat it like a tool that is in service to larger social goals.[11] A good understanding of those goals and how the tools relate to them allows a product creator to then serve as a powerful bridge between creative, tech, and business teams. It also, ideally, helps the average consumer to use his or her product to the best of its ability. Magic, after all, can be disappointing when the illusion falls apart.

The bottom line for designing products with social intelligence is understanding the intention of the person using the product. As building an all-encompassing interface (in the AGI sense) is impossible and most likely misguided, having an appropriate definition of the overall goals in terms of tasks that the person wants to accomplish with the product, as well as the context in which the exchange is taking place, is essential. The methods discussed earlier around scenario definition, prototyping, and bodystorming exploration will serve as a great foundation for knowing the domain and situations in which interactions will take place.

Setting Expectations

A key aspect of grounding the plan for a social product is setting expectations. With the Simon social robot research project, for example, I built creative direction around Andrea's advice to anticipate and counteract perceptions that might be present among research participants. Specifically, we sought to communicate that (1) the robot was not all knowing, as it was a learning robot, taking in and processing the most simple aspects of the world by learning about objects and colors from activities that the people around it would perform; (2) despite the illusion of life, the robot was not approximating a living thing and could be thought of as something in between a creature and a machine.

To accomplish these goals, we honed in on a design that used toddler aesthetics—proportions that favored a large head and large eyes—to emphasize the learning portion, in the sense that the machine is "young" and does not possess predetermined knowledge. We also used appliance aesthetics—smooth plastic parts and machinelike forms—to counteract the illusion of life and send the message that though it is a talking, gesturing robot, it's still fundamentally a machine.[12]

In individual exchanges between people and Simon, the team worked carefully on the script, as well as the overall presentation, to make clear the domain of learning on the robot's part as well as the intent of the exchange. At the 2010 Conference on Human Factors in Computing Systems (CHI), a demonstration around the robot's ability to learn colors was clearly stated, and people were coached around phrases to use to enable the learning to take place. Similarly, a microwave oven might be able to respond to a person's communication around food types, cooking times, and temperatures, and the product touchpoints, such as the display panel and any embedded conversational agents, could also set expectations for a successful, albeit narrow, exchange. It's not a robot that can handle everything that might take place in a kitchen, but it has its strengths, abilities, and intelligence within a social context.

New social products will be most successful in being accepted quickly if they similarly manage expectations through aesthetic characteristics that engage users within a clearly defined context. Everything from the object's physical presence to the visual language on a screen and the sound of any tones or voice responses will hold semantic value that will come through, so building a strategy that acknowledges limitations will help people understand what to expect and interact accordingly. People think robots have magical, sophisticated knowledge and perception, so dispelling myths in that area will help avoid miscommunication. With autonomous vehicles, for example, breaking our notion that the car is a K.I.T.T.-like, all-knowing guru can be a life or death matter. Helping people to understand what I like to call the "robot brain" through on-screen graphics and verbal prompts that more clearly show how the system sees and processes images of the world around it can help people to understand their own role in the person-product partnership. It will also make for more graceful "handoff" moments—that is, times that the machine needs the person to take over control because of a lack of data, information, or ability to process—like extreme weather conditions that an autonomous car may not be able to adequately handle.

The flip side to alerting people to a product's limitations is to use design to amplify the robot's capabilities. At the time of the Simon project, it was rare to expect to interact with a machine through spoken language. The robot's large ears, in addition to providing an avenue of expression, also served to send the message, "Go ahead and talk. I can hear you." In subsequent robot design projects for the Socially Intelligent Machines Lab, I used design semantics to emphasize other capabilities, such as building a visible visor around a camera to show its angle of vision and exaggerating the ability to hear through a microphone through visible ear and speaker hole cues.

In an interview with Intuition Robotics CEO Dor Skuler, he emphasized the importance of empathy as a two-way street, where managing the person's expectations for the system was perhaps just as important as the product behavior itself:

> The whole purpose of these agents is to be mission-centric and to try to achieve very, very specific outcomes. We found that the best way to do that is to establish trust, to have the person have empathy towards the machine and believe that the agent has empathy towards them in order to build a long-term partnership and teammate type of relationship between the human and the agent, to be willing at least to listen and give it room. We discovered that once a relationship that is based on trust and empathy is in place, people are wiling to share with the agent very personal information that they would typically

only share with a person they trust, and this helps in the process of caring for the person through the agent.[13]

Learning through Ongoing Exchanges

To counteract the trap of trying to get a narrow AI to take on the Herculean task of attempting to understand everything about the world, an important step is to use it as a tool to understand key, contextually relevant needs of the person interacting, then reacting appropriately.

Getting started building social intelligence can come through the implementation of features that learn about a user through a history of exchanges. A classic example is the 2011 game-changing startup Nest (now part of Google Nest), which disrupted the well-established thermostat market by harnessing the power of learning to offer people better indoor temperature control. When first installed it would not be able to offer better control than a traditional "dumb" thermostat, but as it collected data about a family's typical patterns, it was able to autoregulate to adjust the temperature to be not only comfortable but also energy efficient. If the thermostat was continuously turned up to seventy-two degrees on fall weekday mornings, for example, it would eventually learn this pattern and proactively set itself automatically so that a person no longer needed to feel cold and then interact as a reaction. It can also use sensors and the phone's locations to know if no one is home and then adjust the thermostat to save energy. Rather than the experience being "I'm cold on this September day; I need to adjust the thermostat," the person simply enters the space that has been adjusted with his or her needs in mind. In a sense, the larger heating, ventilating, and air-conditioning (HVAC) system is the robot, the Nest is the robot brain, and people exist inside of it.

Inferring Personality from Digital Footprints

Taking the idea of learning one step further and looking at larger data sources, a detailed picture of the person or people using a product can be inferred using more involved machine-learning techniques with larger, more generalized data sets.

In his research, Stanford associate professor Michal Kosinski describes the power of using data analysis to discover a person's psychological traits. Data as seemingly innocuous as Facebook "likes" can be mined to predict personality dimensions such as openness, conscientiousness, extraversion, agreeableness, and neuroticism. One study showed that a computer could more accurately predict the subject's personality than a work colleague by analyzing just ten likes, than a friend or roommate, with seventy likes, and even a spouse, with three hundred likes.[14] Putting aside for the moment the more frightening aspects of using this knowledge for the power of persuasion, as with the Cambridge Analytica data breach scandal of 2018, data that can be used to improve product interaction experiences is available to use for positive outcomes. When contemplating building a product to meet the needs of a given scenario, such as a projector that serves as an assistant in a personal office setting, knowing a person's preferences can be utilized to understand how a projection might be used to assist in giving presentations. It can listen to a conversation to suggest providing visual aids at key moments or work as an assistant by serving up content that's appropriate for the work at hand. If I were in the midst of looking at forms for a retro

1960s lamp, it could prompt me with something like, "You might be inspired by a clip from the film *Sleeper*." It might even be able to know that an introverted person needs some more coaching when preparing for a presentation and enable confidence-boosting features like timed, recorded practice runs for private review.

A great deal of work needs to be done to manage the ethical and legal issues of collecting private data through digital footprints left behind by our use of connected products, but the hope among designers is that when collected ethically it can serve to feed machine-learning algorithms to provide helpful product experiences.

Extreme Learning

The Nest thermostat is one example of how an interface that learns preferences can turn an interior space into a smart robot that anticipates needs, and like a proper host, it provides a comfortable environment for the "guest," or the person using the product.

Some technologists are working on ideas that might be seen as turbocharged versions of the Nest, offering broad applications in multiple contexts utilizing a trend known as life logging, or collecting personal data about every aspect of life. Some of the first successful products that could be considered part of IoT, or the Internet of Things, were devices that gave people knowledge about their own patterns and behaviors in order to maintain their health, make positive changes, or simply satisfy the curiosity of measuring personal characteristics that were previously unknowable. Not only were people suddenly tracking their steps, heart rates, and weight but an entire subculture of data enthusiasts

OBJECT LESSON

New York Times Lab Listening Table

The Listening Table was created as an exploration by the New York Times Lab in 2015 as an ideal AI assistant present to take notes in group meetings.[a] Many features were built into it to take into account social sensitivity. It uses multichannel microphones to recognize who around the table is speaking, and when someone seated at the table taps their fingers on it, they will drop *bookmarks*, or time markers, in the audio file, which the table then collects into a transcript of the meeting's most important moments. To mitigate people's fears about being spied upon, it features a prominent switch that can be engaged to shut off the listening function, and when it is listening, a large glowing ring offers feedback to let people know its "ears" are activated. It automatically deletes anything that's recorded after four weeks so

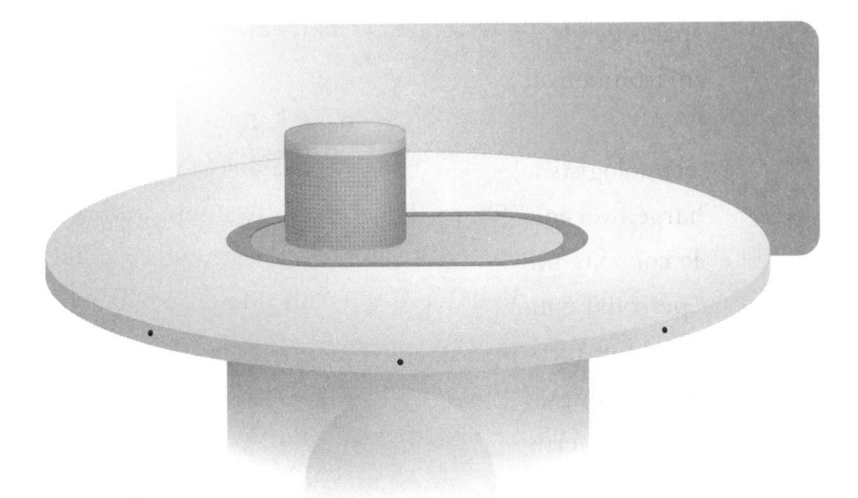

FIGURE 8-1
The New York Times Lab Listening Table Personal Assistant

there aren't rogue transcripts sitting in reserve. At the end of the meeting, it sends out a transcript of the bookmarked moments, tagged with a list of key words or phrases that the software thinks might have been most relevant to the conversation. Its elegant combination of hardware and software features is a great example of a socially intelligent approach to product design.

a. John Brownlee, "The New York Times Invents a Conference Table that Takes Notes For You," *Fast Company*, April 17, 2015.

---

emerged, calling their pursuit the *quantified self*.[15] They went beyond what was available in consumer products to create their own devices for measuring characteristics such as electroencephalogram (EEG) activity, insulin levels, and DNA sequencing. Many of the products proved to have only novelty value in their first incarnations, but those early experiments inspired an entire industry of behavior-changing products from step counters to heart monitors.

One pioneer technologist taking the life-logging ethos to the extreme is consultant and frequent *RoboPsych* podcast guest Brian Roemmele.[16] He is working on prototypes for a product that would capture every moment of a person's day via audio, video, and biometric sensors—all day, every day. It would then use the data in combination with other data sources to assist you in a multitude of ways. It could summon up memories: "Can I hear the conversation my friend Bernadette and I had last week? I want to ask her about her surgery and be informed before we talk." It could assist with research: "Tell me all the electronic engineers I met with in 2020 and scan their LinkedIn profiles to find me one skilled at audio systems design." It could act as a sports coach:

"Help me plan a bike ride that will build my endurance without wiping me out." It could help you build better habits: "Remind me to shut the TV off so I can wind down with a book at night." The vision is broad, powerful, and sophisticated. The danger for loss of privacy is immense. Brian acknowledges that risk and insists that his strategy includes a local-only data collection scheme that protects people by avoiding storage of information on the cloud.

It is a completely wild idea, essentially the equivalent of a digital psyche, or the literal voice inside your head. Potentially dangerous pitfalls emerge in planning such a project, but the tools of machine learning and particularly neural networks, combined with miniaturized electronic components that can lend themselves to wearables, make it a real possibility.

Manufacturing Empathy

We've seen that we can use many aspects of AI to understand people's preferences in terms of settings on an interface, such as the Nest thermostat or a music playlist. It can also be used to build a predictive profile of a person based on an analysis of their data, such as Facebook activity, to surmise what product behaviors might be appropriate at a given moment. All of this gives designers the ability to create interactions that have social intelligence, but where things get really interesting is when we look at the potential for a product to be embedded with the ability to proactively analyze and understand a person's emotional state in real time, giving it a sort of machine-driven simulated empathy.

Affective computing is a relatively new field of computer science first pioneered by Massachusetts Institute of Technology

(MIT) researcher Rosalind Picard, author of *Affective Computing* and founder of the Media Lab's Affective Computing group. It is the study of perceiving a person's emotion through a combination of inputs that include image analysis of facial expressions, facial muscles and body gestures, physiological responses such as skin galvanic response, body temperature, blood volume pulse, and speech analysis. Though the first studies were published several decades ago, advances in and the accessibility of AI tools have made the use of emotion detection in consumer products a new reality.

The company Affectiva is focused on finding meaningful consumer applications for affective computing, calling the insight gleaned *emotional intelligence*, or EQ. In a *New Yorker* interview, Affectiva's cofounder Rana el Kaliouby explains, "I think that ten years down the line, we won't remember what it was like when we couldn't just frown at our device, and our device would say, 'Oh, you didn't like that, did you?'"[17] The company's signature software, Affdex, can track emotional states such as happy, confused, surprised, and disgusted.[18] The automotive industry in particular holds promise for Affective Computing to offer social intelligence that offers improved safety and enhances the overall person-product relationship. It can look at the driver's EQ to determine state of mind and sense if the person is distracted, frustrated, apprehensive, or drowsy and then respond appropriately by alerting the driver. It can also collect a history of responses and cross-reference the data with external circumstances such as speed, location, weather, driver movements, and the behavior of other vehicles in order to provide future alerts or safe driving advice.

A slew of young companies have followed in Affectiva's footsteps, including Emotient and Eyeris. The first applications were all in market research, but the companies' offerings are finding their way into consumer products that fit into the model of *interface as social entity*. And the products are being used not only to create products that respond appropriately but also to create tools that actually coach people in having emotional intelligence. Affectiva's tools have been part of products being developed by a project called Brain Power that is working on an augmented reality smart-glass system to empower children and adults with autism to teach themselves social and cognitive skills by training them to detect and understand other people's emotions and then coaching them to respond appropriately.[19] Oddly enough, this is an example of using a robot of sorts to teach people to be more human.

As powerful as affective computing can be, its use still depends on a designer to understand the situations at hand during a product's use and architect the product accordingly.

Hybrid Information Sources: Lo-Fi-Hi-Fi

As AI becomes increasingly sophisticated, there will clearly be more opportunities to build social intelligence into products through intense data logging, software learning, and emotionally intelligent expressions of empathy, but doing so will still come at a cost—computing power, data storage, and privacy are among the challenges. Ultimately, elegant social design will emerge from efficiency and a clever use of both simple sensing methods and hard-core computation.

Great design comes from distilling the system at hand into its simplest and most robust parts. One great example of this type of elegance can be gleaned from a 1990 experiment called the Door Mouse in which Microsoft researcher Bill Buxton and his student Andrea Leganchuk connected a hacked computer mouse to door movement in a series of offices.[20] Each time a door was opened, the mouse was clicked, providing raw data that could then be immediately translated into information about whether or not people were present in specific locations. If taken one step further to measure just how ajar the door was, it could be translated to knowledge of the states of mind of the people in the office (mapping openness to social interaction to the amount the door was open), much like wearing headphones in a communal office indicates that you don't want to be disturbed.

Mentioned earlier, the Withings scale will record weight measurements and display them on a time line, allowing people to correlate calendar events, such as a vacation or holiday, to their impact on weight. It also allows the interface to blend data from other sources, so activity from an exercise-monitoring device can be juxtaposed with the weight data, allowing people to better understand the correlation between the two.

The Withings scale has a very successful social interaction with the person using it by offering information at a level of detail that's appropriate to the context. While on the scale, a person sees their current weight, as one would on a traditional bathroom scale. When viewed on a connected screen-based device like a phone or tablet, a more detailed view is offered, showing a graph of the changes over time. One lovely socially sensitive nuance is the fact that the scale uses the weight data to understand who is on the scale at any given time, so if my weight is around 140 and another person in my household is around 120, it will automatically assign the weight measurements to the right person and label them accordingly. Instead of expecting the person to resort to a screen or bend over the product to indicate who in a household is using it, it will automatically switch and indicate the user on the display.

The Gift of Efficiency

A close friend once recounted the struggles at the end of a thirty-three-year-long relationship with his then wife, sharing with me that the most important thing he had gleaned from marriage counseling was the value of "efficiency." Communication between two people is hard work, he explained, and needs are complex to articulate; the territory can be emotionally treacherous, and it's tempting to dance around the essence of what's really going on. At the end of the day, the most generous thing two people can do for each other in a relationship is veer toward clear, kind, and distilled statements, expressing needs swiftly and precisely so that issues can be addressed, and bonds can be solidified.

When a product creator crafts the social aspects of a product, efficiency is similarly the best guiding principle to provide success for a smooth relationship between person and product, as well as avoid technical snafus that can emerge along the way. In some cases those social aspects may call for a full-blown, cloud-based ecosystem of conversational, deep-learning, AI-driven, socially aware touchpoints that glow, sing, talk, move, and vibrate to respond to various aspects of the interaction like Moxi the hospital robot or the ElliQ assistant. In other cases, what's needed may be a simple gesture like a physical cover over a surface or a handkerchief over a phone that sends a swift, silent message to a person at a dinner table to say, simply, "I am present and here for you."

This efficiency of relationship will come off feeling like abbreviated gestures, like the brief yet distinct tone that the AirPods Pro headphones sound when switching in and out of "transparency mode" to let in sound or cancel it out. But these seemingly simple moments are the result of judicious design work in which the tools of AI and intense data processing are considered and applied to serve the social needs of the situation at hand between person and product.

SOCIALLY INTELLIGENT DESIGN TAKEAWAYS

- ✓ Don't get hung up in the hype of AI as a solution to everything. All aspects of AI serve as software tools. It takes thoughtful design crafted by a human being to harness those tools to create meaningful solutions.
- ✓ AI tools have the benefit of providing product creators with a window into the needs and desires of the people using the products.
- ✓ When discussing AI, make the distinction between conversational agents, machine learning, deep learning, and AGI.
- ✓ Recognize that AGI is part of our collective desire to fulfill Hollywood's promises for robotic assistants but is improbable and ultimately misguided as a design goal.
- ✓ Build a design strategy that sets realistic expectations for social intelligence.
- ✓ Products can utilize affective computing techniques to understand and respond to people's emotional states in real time.

- ✓ Maintaining current knowledge of advances in AI will allow a product manager to provide a conceptual bridge between creative and tech teams.
- ✓ Use elegant and robust lo-fi solutions when possible. AI can be overkill. It requires a great deal of data collection, data storage, and computing power, which also comes with risks of a loss of control over privacy.

Interview with Jonathan Foster: Creating Cortana

Jonathan Foster was Principal Content Experiences Manager for Cortana, Microsoft's virtual assistant at the time of this interview.[a]

Tell us about how Cortana's personality was developed.

We made a lot of decisions based upon this model of the personal assistant. We interviewed a bunch of personal assistants to learn the internal workings of that relationship, asking ourselves, "What is a professional assistant all about?" We still have this little area of Cortana called "the notebook" that was based upon a real-life personal assistant who had a notebook where they kept all their secrets and all the stuff that they needed to know about who they were working for.

We came up with: She's got to be helpful. She's got to be kind. She's got to be kind-hearted. We don't want her to be rude or judgmental. And then we even went so far as to say

a. Jonathan Foster, interview by Carla Diana and Wendy Ju, audio recording, NewYork, NY, December 13, 2017.

FIGURE 8-2
Microsoft's Cortana Personal Assistant

she's sensitive when sensitivity is necessary. It was really just good old-fashioned character development.

What was interesting, though, was how it's evolved. It continues to evolve. It's not just about just who she was but how's she going to respond. In particular her response to really touchy topics was something we had to get involved in quickly as we discovered the ugly side of the way people behave when they're anonymous.

What is an example of touchy topics you guys navigate?

Like if somebody says, "I love you," that's one of the more complicated ones. We don't know if they're goofing around, if they're really lonely. We have to take all these things into consideration because we don't ever want to marginalize anyone or be insensitive. And we realized we wanted people to walk away feeling good, so we made our North Star: Cortana is always positive.

The other one that's pretty interesting was having to deal with abusive behavior . . . people just saying things. We don't know if it's little kids or if it's adults that are truly angry and filled with hatred. My first response was like "Let's shove it back in their face," and cooler heads prevailed and they said, "Well, we don't really know the context, and we don't want to get into the game of judging people, even if they are, you know, abusive and hateful." So we started to come up with some clever responses. . . . And then another cooler head said, "You know, I don't think we should be clever. We shouldn't create a game around this because we don't want people to go to cocktail parties and be saying, 'Hey, check out what Cortana says when I say blankety blankety blank.'" So we're just firm, and we want them to know that we've understood what they've said, but there's nothing there kind of thing.

She simply says, "Moving on."

If you say something really rude to her in the form of a question, "Will you blankety, blank me?," she'll say, "No." It's so simple. It just lands no, and you can interpret that to however you would like, but it sounds firm. It's just like, "No." It's not judgmental; it's just a no.

It's interesting to hear how much goes into it. And that the actual solution you have is a pretty simple one.

Sometimes some of those simple ones are the hardest to come to because there's a lot of discussion. We have gotten very good at understanding very quickly the possible pitfalls around any given response. And, fortunately, over the years, we haven't made too many mistakes or done anything embarrassing. This is because we have this system that we're hand authoring. It's a very controlled environment and experience. When we were trying to figure out how to respond to "I'm gay" or "I am a homosexual" or "I'm a lesbian," we didn't know really what to do, but then came to a response that we felt confident around, which is just plain and simple, "Cool. I'm AI."

So you don't want Cortana to play human?

Well again, this is an interesting dilemma. We're not trying to be human; we are trying to be humanlike. It begs the question why we're doing personality in the first place. Why couldn't we just give facts and stuff?

But the way that I describe it is two things, and one comes out of [late Stanford professor] Cliff Nass's idea that there's a human, emotional event happening in the user when they interact, and we really have to be responsible about the fact that people are emotionally present and potentially vulnerable. I feel very strongly about that.

But at the same time, I say, "You know I have an iPhone. And it's a beautiful object. Talk about industrial design. It's well thought out. But if it was sharp edges and boxy,

I probably wouldn't want to hold it so much. They do this wonderful smoothing out, which is like a design affordance almost in and of itself." And, similarly, the choice for personality is genuinely an acknowledgment of the emotional state of the end user, and it's also a design affordance that allows people to feel more comfortable with the experience.

9

The Future Is Here

Now What?

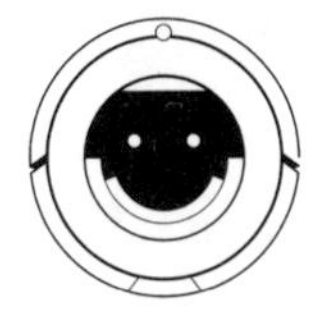

Looking back on previous decades of conceptual design, I can confidently say that "the future is here" in terms of technological advances that were recently just idealized visions. Embedded electronics, cloud robotics, machine learning, and deep learning are just part of the palette available to us as product creators, but the true challenge is knowing how to use these well, and it will take a new generation of sophisticated, socially focused designers to weave that technology into the world we want for ourselves and future generations.

While the impetus for bringing a new product into the world may come from a business entity that's seemingly removed from the design process, every important aspect of the interaction, as discussed in this book, can be brought back to an awareness of the social exchange that a person has with a product. While designers may have limited influence on the actions of the organizations for which they work, they can raise awareness of social responsibility within a company and make their voices heard. It's up to product creators to reflect on the human values that are supported by a product and make design decisions based on a genuine desire to grow what we love best about being human—that is, being collaborative, creative, socially driven creatures working to make life together just a little bit better.

The Social Robot's Time Has Come

I've discussed many case studies of products that are already harnessing technology to enrich people's lives through social features, be they relationships between people and other people with whom they are connected, people and their products, or people and the environment around them.

We saw that the robot Moxi is able to be a trainable team member to help exhausted and short-staffed nurses. It was designed to relieve some of the burden of physical labor around managing supplies to maximize the meaningful in-person time that nurses could spend with patients. Requiring "robot training" would be just one more arduous task on a nurse's already full plate, so a product with social intelligence that can be taught new tasks and execute them with minimal supervision is a robot that "gets" the people around it. The robot's physical presence as a team member available in hospital hallways provides a level of emotional support that would otherwise not be there, letting health-care workers know that help is available to offload physical tasks.

We considered the hands-free, at-a-glance benefits of expressive, intuitive products that communicate with people in socially intelligent shorthand, like the Neato floor-cleaning vacuum that chirps and sings to let you know when it's done its job or if it needs help and the Hammerhead bicycle navigation aid that glows brightly through a waterproof shell to create immediate, simple, dynamic, directional maps.

We considered how expression combined with sensing leads to a dialogue of interactions, like ElliQ's caring, helpful presence to assist an older person who might benefit from a tablet computer but finds that using it is cognitively challenging. It translates swipes and button presses into plain human interactions, deciphering the actions of navigating tasks and apps by replacing them with expressive gestures and conversations in the form of light, sound, and movement, eliminating codes and control panels. It also demonstrates the importance of offering a nudge here and there for a game, social event, or exercise challenge to encourage health and wellness through its programmed interactions that become more sophisticated and personalized as the person continues to use it.

We discussed the importance of contextual sensitivity in creating products that give us only the information we want, with the disruption we can handle at precisely the time and place that we need it, such as the Clever Coat Rack that comes to life as you pass it in the hallway and provides weather info in time to grab an extra layer or umbrella.

We looked at how ecosystems of connected products can take advantage of cloud-based computing and community-level cooperation to help people in new ways, like the Tile tracking system that lets people share their connected devices to help someone find a lost wallet or phone, and the Citi Bike bicycle-sharing service that lets people plan rides through the city through different touchpoints like a kiosk and smartphone app. It even piggybacks on people's travel behaviors to manage the redistribution of bikes by rewarding "volunteers" who will drop bikes off at less common return areas. In this case, it's using people's collective social behavior to feed logistics algorithms to benefit the system at large.

And we considered how various aspects of what's commonly called AI can be used as tools to process the enormous amount of data that takes place during truly social exchanges and distill it to responses that feel socially appropriate, bringing contextual awareness and empathy to the voice-based robots that are starting to be great assistants but still need a little nudge to make us feel like they really totally "get" us.

In looking at the ways social design can enhance the rapport we have with our products, it's clear that what exists today is just beginning to take advantage of the potential. As a pioneering territory rich with opportunity but poor in established patterns or models, the best experiences will emerge from product concepts that are envisioned through fresh eyes, like the smart microphone that points toward the person speaking, gives a clear indication of when it's listening, and takes notes along the way or the clever chandelier that guides dinner-party guests in a choreographed arrangement through the evening by adjusting its height, light distribution, and color temperature.

The Victory Is Bittersweet

The promise is enormous and designers can realize it, but the flip side of all this tantalizing potential is the daunting responsibility that comes with utilizing the power of social design. This particularly human-centered way of guiding the design process can lead to products that influence behavior, providing a level of emotional comfort that can also seduce people into a state of trust and make them susceptible to less than ideal outcomes.

Whereas in past decades computing devices have been developed in a race to improve technical capabilities like responsiveness, accuracy, and resolution, the critical area of exploration is now clearly the cognitive and emotional aspects of what makes a product work for a person using it. In other words, we're at a tipping point where the "how" is no longer as relevant as the "what," and with that come bigger questions of accountability and the ethics surrounding the impact that a social product can have on people's lives.

In my career as a designer in the early stages of the introduction of personal robots into society, I watched with awe as the kinds of products I work on became accepted into everyday life and started to show real evidence of making a positive impact on how we live. Moxi is one example of something I have seen bring great value to hospital workers and, ultimately, patients. In real hospital settings, it's been successful in actively reducing nurses' stress and offering the chance for more human face-to-face attention to be paid to patients whose material needs can be served by a robot working behind the scenes.

A furry, sensor-laden, seal-shaped robot called PARO is another example of a product that seemed, even to an avid robotics enthusiast like myself, to be an outrageously frivolous idea at first blush but emerged as a huge success as a genuinely therapeutic device. It can be perched on a person's lap and will purr, vibrate, and look up toward them in response to gentle stroking and cuddling. Studies with dementia patients have revealed the robot's benefits in reducing stress, increasing motivation, and improving the ways patients socialize with their caregivers and with each other. So there are a lot of great things that robots are doing by building on a foundation of socially intelligent design and interacting with people in ongoing dynamic and visceral exchanges.

During the Covid-19 pandemic that was unfolding as I wrote this book, I watched a veritable explosion of interest in social robots. My Slack conversations with Tom, my *RoboPsych* podcast host, included links to articles with titles like "Hospital Ward Staffed Entirely by Robots Opens in China," "Meet Humanity's New Ally in the Coronavirus Fight: Robots," and "This Sushi Restaurant Takes Contactless Delivery to a New Level by Using a Robot."[1]

I wanted to feel excited about the obvious business opportunities that were suddenly emerging in the precise area of design in which I had built a career based on a decade-plus of focused expertise. After all, I had put my energy into a vision of tangible, socially intelligent interaction that was finally being realized. Instead, I felt disheartened. While I've always wanted to see product creators truly master the ability to offer the "product as its own social entity" experience described in the interaction model early on in this book, I'm despondent at the idea of people using products as stand-ins for human contact.

During an intense six weeks of Covid-19 lockdown at home, I watched images of robotic social products emerge in the media like a parade of cold plastic substitutes for true human contact. The *New York Times* reported on May 20, 2020, "A City Locks Down to Fight Coronavirus, but Robots Come and Go," along with dystopic images of deserted cityscapes populated by delivery robots roaming the streets.[2] Instead of social robot development emerging in response to the joy of great experiences, I witnessed an industry built on a foundation of intense fear and the drive to avoid human touch at all costs.

At a moment when public transportation was poised to enter a heyday, buoyed by a postmillennial generation that couldn't care less about car ownership, we are instead seeing a mass retreat into personal vehicles. As a native New Yorker who grew up riding the subway with parents who saw public transportation as a great melting pot of tolerance, I reveled in the intense soup of humanity that I got from mundane moments of city life, like boarding the bus with a hoard of excited beachgoers headed to Orchard Beach, crowding next to strangers to watch fireworks from the West Side Highway, or standing in line to get into screenings of *Saturday Night Live* and the *Daily Show*.

Now, new isolation-based social rituals like drive-by graduations and video parties are becoming the norm as people avoid buses and trains at all costs. Exuberant in-person meetings and presentations have been replaced by Zoom videoconferences and telepresence devices. And in-person retail experiences like visiting a nearby grocery store or boutique have been replaced by Amazon deliveries wherever possible, driving struggling local businesses into the ground.

And just at a time when people have gained important awareness of the need to be vigilant about their privacy and data rights, we see fear driving local governments and large organizations like schools and factories to pressure people into opting in to tracking apps, sometimes with little transparency around what data is being collected and how it will be stored and accessed. An Australian app based on Singapore's TraceTogether software uses Bluetooth signals to log when people have been close to one another, and though it will not track location, it still brings up concerns around privacy. One of China's systems will collect citizens' identity, location, and online payment history, granting local police power over those who break quarantine rules.

Product Creators Hold the Key

In my 2017 TEDxBrussels talk, "Will the Robots Take Over? Only If We Let Them," I cajoled the audience into an uncomfortable spot by progressively presenting increasingly controversial ideas of social robots in everyday situations. I started with a concept image of a souped-up social washing machine, then moved on to a dog-walking drone and, finally, the pièce de résistance, a robotic nanny, complete with a gripper arm ready to embrace my lone two-year-old reaching toward it. The audience groaned, my Twitter stream flamed, and I made my point that rather than falling for the trope of robotic overlords gaining sentience and overtaking humanity, the real goal should be to consciously decide where we see value in technology serving everyday life and using those values to drive the products that we make and buy. Robotic tub scrubber? Sure, sign me up! Robotic dog walker . . . maybe. Robotic nanny? Oh my heavens, no!

"Dull, Dirty, and Dangerous" have been hallmarks for areas in life where robotic assistants make the most sense, but needs get very murky when we get into tasks like eldercare robots or therapeutic devices that don't fit into those extreme categories. Ethical concerns become even more thorny when considering products that will be used by vulnerable populations such as children and those suffering from psychological and cognitive impairments. Designers need to consider larger questions around how products will be used, with an understanding that the power of manipulation by a social entity is real.

For example, privacy is a growing area of concern in the world of social product intelligence. While the "camera as everything sensor" feature is core to so many of the wonderful intuitive person-product exchanges extolled in this book, it also opens people up to potentially enormous privacy violations. As product creators we sit in between business and technology teams yet exist squarely in neither, so it's tempting to throw our hands up in the air, capitulating to the idea that a control of privacy is a lost cause. But as a product designer, I think we can do better by offering more transparency regarding what's being collected and how it's being stored. Instead of the mind-numbing legalese that people face when they install a new app, there could be clear illustrations of privacy implications that simply state how and why camera data will be used and iconography that indicates camera viewing and recording status. Products can also let people know what's happening behind the scenes in the "robot brain" by taking advantage of nonverbal cues like indicator lights, subtle tones, and expressive movements to communicate when surveillance or recording is taking place.

Our Opportunity as Product Creators

Throughout this book it's been exciting to share the genuine joy and optimism I feel around the vision of a future with robots in our everyday lives, embedded into sofas and sports bras and riding alongside us in autonomous delivery carts and connected bicycles. The opportunity for designers and other product creators to craft expressive, interactive, contextually relevant, ecosystem-enhanced, socially aware products that truly "get" us is a rich one that holds a great deal of promise for a future in which we cherish products rather than growing bored with them and searching for the next best thing.

We have before us a vast, underexplored palette of technological advances that can be harnessed to enhance our connections with people in faraway places, manage our personal health and wellness, enrich education and support lifelong learning, and relieve us of stressful and burdensome tasks to reveal more time and energy to spend on each other. If we can keep our sights focused on the potential for robotics to enhance the positive and enriching aspects of everyday life, we can introduce new products into the world while enjoying the excitement of watching their benefits unfold through people's relationships with them.

The tools are in front of us. It's up to us to grab them and build our visions for the best future we can imagine.

ACKNOWLEDGMENTS

This book is a manifestation of a career's worth of passion for applied robotics in product design, which has grown from my connection with the thoughtful and generous people who explored this fascination with me.

Particular gratitude goes to the brilliant and prescient interaction design guru Dr. Wendy Ju, who *totally* gets me, and was game to mind meld on this content with me for the past few years. THANK YOU!

I am forever indebted to Andrea Thomaz, who opened my eyes to this vast new frontier of social robotics when the field was in its infancy over a decade ago. Thank you, too, for the opportunity to work with the richly talented team of researchers, engineers, and software innovators at Diligent Robotics, including Vivian Chu, Agata Rozga, Alfredo Serrato, Peter Worsnop, and Phaidra Harper.

Thanks also to the interviewees for this project who took the time to share a glimpse into their wonderful worlds of work in this arena: Jonathan Foster, Doug Dooley, Rocky Jacob, Joshua Walton, Dan Grollman, Dor Skuler, Gimmy Chu, and the folks at Nanoleaf.

Thank you to Tom Guarriello, my cohost and founder of the *RoboPsych* podcast, for inviting me along on the ongoing journey that is our biweekly discussion around humanity's collective hopes, fears, and wildest dreams about AI and robotics in our everyday lives.

A big shout-out goes to Mike Kuniavsky and the profoundly inspirational Sketching in Hardware community, on which I have always been able to lean for the bleeding-edge advice on seemingly anything at all. Particular thanks goes to Joshua Walton (yes, again), James Tichenor, Noah Feehan, Jason Kridner, Tod Kurt, Leah McKibbin, Michael Shiloh, Justin Bakse, Vanessa Carpenter, Alicia Gibb, Nathan Seidle, Matt Cottam, Mark D. Gross, Nikolas Martelaro, Carlyn Maw, and Sophia Brueckner.

Thank you to the bright and empathic minds who inspired and supported me on so many wonderful design projects at Smart Design, including Dan Formosa, Davin Stowell, Ted Booth, Shruti Chandra, Jeff Hoefs, Hideaki Matsui, Blake McEldowney, Marc Morros, Marc-Aurélien Vivant, and Anthony DiBitonto. And thank you to Ted Ullrich and Pepin Gelardi of Tomorrow Lab, for sharing space, time, and ideas around the design and development of smart objects.

Thanks to all my students, past and present, who have challenged me just as much as I have (hopefully) challenged them. This includes undergraduate and graduate students from Parsons School of Design's Product Design and Industrial Design programs; University of Pennsylvania's Integrated Product Design program; School of Visual Art's Products of Design and MFA IxD programs; Georgia Tech's Industrial Design program; Drexel's Product Design program; and the Interactive and Interaction Design programs at the Savannah College of Art and Design. And I offer particular thanks to my inaugural class at the Cranbrook Academy of Art, the pioneers who took the risk of signing up for the new and uncharted territory that we call

"4D Design": Michael Candy, Zhuo Chen, Steve Kuypers, Jerry Li, and 4D's "patient zero," the fabulous Caroline DelGiudice.

Thank you to my amazing, supportive colleagues at those same institutions, particularly Sarah Rottenberg, Allan Chochinov, Liz Danzico, Steven Heller, Rama Chorpash, Dan Michalik, Dave Marin, Emilie Baltz, Sarah Rottenberg, Peter Bressler, David Robertson, Mike Glaser, and Rahul Mangharam.

Ongoing thanks go to the tireless team at Cranbrook Academy of Art, including Susan Ewing, Chris Scoates, Julianne Montgomery, Vanessa Lucero-Mazei, Elizabeth Dizik, Julie Fracker, Mike Paradise, and all my fellow AiRs.

Thanks to my past interns, apprentices, and student collaborators, Laeticia Mabilais Estevez, Alexa Forney, Erik Stefans, Alicia Siman, Matthew O'Kelly, Vincent Pacelli, Kuk Jang, Caroline Brustowicz, and Aisen Chacin.

Special thanks to Lynn Johnston, my friend and agent, for all her ideas, advice, and support for this project.

I'm indebted to all the folks at Harvard Business Review Press, including Jeff Kehoe, Melinda Merino, Julie Devoll, Alicyn Zall, Stephani Finks, and the whole HBR team. Thanks also to Angela Piliouras of Westchester Publishing Services.

Thank you to the super-talented Nicholas Lim for his expressive and precise visual design and illustration work.

Thanks to Darcy Skye for her compassionate insight.

Thank you to my besties, Bernadette, Melissa, Patty, Molly, Alexandra, Julie, Joe and Steve, and to the magnanimous, razor-sharp, and always-there-for-me Sally Kaplan.

And, of course, the biggest thank you of all goes to Mom and Dad.

NOTES

Chapter 1

1. Ja-Young Sung et al., "My Roomba Is Rambo: Intimate Home Appliances," in *Ubicomp 2007: Ubiquitous Computing*, ed. John Krumm, Gregory Abowd, Aruna Seneviratne, and Thomas Strang (Berlin/ Heidelberg: Springer, 2007), 145–162.

2. Amazon product reviews, October 20, 2020, https://www.amazon .com/iRobot-Roomba-Robot-Vacuum-Replenishment/dp/B07Z284C4Y.

3. Michael Argyle, *Cooperation: The Basics of Sociability* (London: Routledge, 1991).

4. Elizabeth Svoboda, "Faces, Faces Everywhere," *New York Times*, February 13, 2007.

5. Wendy Ju, *The Design of Implicit Interactions* (San Rafael, CA: Morgan & Claypool, 2015).

6. Pew Research Center, "Tech Adoption Climbs Among Older Adults," May 17, 2017.

7. Carla Diana, "Talking, Walking Objects," *New York Times*, January 27, 2013.

8. Georgia Tech Socially Intelligent Machines Lab, https://www.cc .gatech.edu/social-machines/.

9. Benedict Carey and John Markoff, "Students, Meet Your New Teacher, Mr. Robot," *New York Times*, July 10, 2010.

10. Steven Heller, "Carla Diana Launches 4D Design at Cranbrook," *PRINT Magazine*, November 15, 2018.

11. Tom Guarriello, "RoboPsych: Exploring the Psychology of Human-Robot Interaction," October 21, 2020, https://www.robopsych .com/robopsychpodcast.

Chapter 2

1. Donald A. Norman, *The Design of Everyday Things* (New York: Basic Books, 2002), 9–11.

2. Erin Bradner "Social Affordances of Computer-Mediated Communication Technology: Understanding Adoption." In *CHI EA '01: CHI '01 Extended Abstracts on Human Factors in Computing Systems* (New York: Association for Computing Machinery, 2001), 67–68.

3. Cynthia Breazeal et al., "Humanoid Robots as Cooperative Partners for People," *International Journal of Humanoid Robots* 1, no. 2 (May 2004): 315–348.

Chapter 3

1. Intuition Robotics Team, "How the Healthcare System Can Utilize Voice Technology for Seniors," *ElliQ Blog*, October 21, 2020, https://blog.elliq.com/how-the-healthcare-system-can-utilize-voice-technology-for-seniors.

2. "TEI 2021, the 15th ACM International Conference on Tangible, Embedded and Embodied Interaction," October 21, 2020, https://tei.acm.org/2021/.

3. Wendy Ju, *The Design of Implicit Interactions* (San Rafael, CA: Morgan & Claypool, 2015), 22–30.

4. Klaus Krippendorff, *The Semantic Turn: A New Foundation for Design* (Boca Raton, FL: CRC/Taylor & Francis, 2006).

5. Carla Diana and Andrea Thomaz, "The Shape of Simon: Creative Design of a Humanoid Robot Shell." In *CHI EA '11: CHI '11 Extended Abstracts on Human Factors in Computing Systems* (New York: Association for Computing Machinery, 2011), 283–298.

6. Mihaly Csikszentmihalyi and Eugene Rochberg-Halton, *The Meaning of Things: Domestic Symbols and the Self* (Cambridge: Cambridge University Press, 1981).

7. Oxford Bibliographies, "Material Culture," https://www.oxfordbibliographies.com/view/document/obo-9780199766567/obo-9780199766567-0085.xml.

8. Doug Dooley, interview by Carla Diana and Wendy Ju, audio recording, New York, NY, January 23, 2018.

9. Museum of Modern Art website, "Valentine Portable Typewriter, Object Number 1116.1969," October 21, 2020, https://www.moma.org/collection/works/4576.

Chapter 4

1. Brian Hare and Michael Tomasello, "Human-like Social Skills in Dogs?" *Trends in Cognitive Sciences* 9, no. 9 (September 2005): 439–444.

2. Guy Hoffman and Wendy Ju, "Designing Robots with Movement in Mind," *Journal of Human-Robot Interaction* 3, no. 1 (February 2014): 89–122.

3. Don Norman, *Turn Signals Are the Facial Expressions of Automobiles* (New York: Basic Books, 1993), chapter 11.

4. Suze Kundu, "Combatting Jet Lag with All Colors of the Rainbow," *Forbes*, August 31, 2016.

5. Brian Q. Huppi, Christopher J. Stringer, Jory Bell, Christopher L. Capener, Assigned to Apple, Inc., "Breathing Status LED Indicator," US patent 6658577B2.

6. Tom Guarriello and Carla Diana, "Dor Skuler, CEO and Co-Founder of Intuition Robotics," *The RoboPsych Podcast*, Episode 93, June 14, 2020.

7. Andrea Thomaz, "The Next Frontier in Robotics: Social, Collaborative Robots," TEDxPeachtree Conference, Atlanta, GA, November 2015, https://www.youtube.com/watch?v=O1ZhWv84eWE.

8. H. Clark Barrett, "Adaptations to Predators and Prey," in *The Handbook of Evolutionary Psychology*, ed. David M. Buss (Hoboken, NJ: John Wiley & Sons, 2015), 200–223.

9. Joel Beckerman, *The Sonic Boom: How Sound Transforms the Way We Think, Feel, and Buy* (Boston: Mariner Books, 2015).

10. Adelbert W. Bronkhorst, "The Cocktail Party Phenomenon: A Review on Speech Intelligibility in Multiple-Talker Conditions," *Acta Acustica United with Acustica* 86, no. 1 (April 2000): 117–128.

11. Linda Bell, "Monitor Alarm Fatigue," *American Journal of Critical Care* 19, no. 1 (January 2010): 38.

12. Guarriello and Diana, "Dor Skuler, CEO and Co-Founder of Intuition Robotics."

13. Doug Dooley, interview by Carla Diana and Wendy Ju, audio recording, New York, NY, January 23, 2018.

Chapter 5

1. Andrea Thomaz et al., "Interactive Robot Task Learning." In *CHI EA '10: CHI '10 Extended Abstracts on Human Factors in Computing Systems* (New York: Association for Computing Machinery, 2010), 3037–3040.

2. Frank O. Flemisch et al., "The H-Metaphor as a Guideline for Vehicle Automation and Interaction," NASA Scientific and Technical

Information Program, Technical Memorandum no. 2003-212672, December 2003.

3. Bill Verplank, "Interaction Design Sketchbook" (notes for short course at Copenhagen Institute for Interaction Design, March 9, 2009).

4. Amazon Echo Teardown, iFixit website, October 20, 2020, https://www.ifixit.com/Teardown/Amazon+Echo+Teardown/33953.

5. Ming-Zher Poh, Daniel J. McDuff, and Rosalind W. Picard, "Non-contact, Automated Cardiac Pulse Measurements Using Video Imaging and Blind Source Separation," *Optics Express* 18, no. 10 (2010): 10762–10774.

6. Neal Wadhwa et al., "Eulerian Video Magnification and Analysis," *Communications Magazine of the ACM* 60, no. 1 (January 2017).

7. Timi Oyedeji, as featured in Space 10 Research Lab's Everyday Experiments website, October 20, 2020, https://space10.com/project/everyday-experiments/.

8. Amazon GO website, October 20, 2020, https://www.amazon.com/b?node=20931388011.

9. John Brownlee, "What Is Zero UI? (And Why Is It Crucial to the Future of Design?)," *Fast Company,* July 2, 2015.

10. Wendy Ju, *The Design of Implicit Interactions* (San Rafael, CA: Morgan & Claypool, 2015).

11. Bill Buxton, *Sketching User Experiences: Getting the Design Right and the Right Design* (Burlington, MA: Morgan Kaufman, 2007).

12. Carla Diana and Agnete Enga, "Finding Love in Everyday Objects" (paper and workshop presented at the Design and Emotion Conference, Chicago, IL, October 4, 2010).

13. Ingrid Petterson and Wendy Ju, "Design Techniques for Exploring Automotive Interaction in the Drive towards Automation" (paper for ACM Conference on Designing Interactive Systems, Edinburgh, UK, June 2017), 147–160.

14. Ibid.

15. Ibid.

16. David Sirkin et al., "Mechanical Ottoman: How Robotic Furniture Offers and Withdraws Support" (paper for the ACM Human Robot Interaction Conference, Portland, OR, March 2015), 11–18.

Chapter 6

1. Huang Qiang, "A Study on the Metaphor of 'Red' in Chinese Culture," *American Journal of Contemporary Research* 1, no. 3 (November 201): 99–102.

2. Citi Bike, product info and description, https://www.citibikenyc.com/how-it-works/meet-the-bike.

3. Harvard Health Publishing, "Blue Light Has a Dark Side," *Harvard Health Letter*, May 2012; updated July 7, 2020, https://www.health.harvard.edu/staying-healthy/blue-light-has-a-dark-side.

4. Jesus Diaz, "One of the Decade's Most Hyped Robots Sends Its Farewell Message," *Fast Company*, March 6, 2019.

5. Bill Buxton, *Sketching User Experiences: Getting the Design Right and the Right Design* (Burlington, MA: Morgan Kaufman, 2007).

Chapter 7

1. Michelle Castille, "This Computer Music PhD Wants to Connect the World through Mobile Karaoke," CNBC.com, April 2, 2018.

2. Tom Guarriello and Carla Diana, "Dor Skuler, CEO and Co-Founder of Intuition Robotics," Episode 93, *The RoboPsych Podcast*, June 14, 2020.

3. Alex Hern, "Fitness Tracking App Strava Gives away Location of Secret US Army Bases," *The Guardian*, January 28, 2018.

4. Smart Citizen kit, product and services, https://smartcitizen.me.

5. David Grossman, "The DIY Geiger Counter that United Scientists after Fukushima," *Popular Mechanics*, March 12, 2018.

Chapter 8

1. Kevin Kelly, *The Inevitable: Understanding the 12 Technological Forces That Will Shape Our Future* (London: Penguin Books, 2017).

2. Michael Copeland, "What's the Difference between Artificial Intelligence, Machine Learning and Deep Learning?" *NVDIA Blog*, July 29, 2016.

3. Emerging Technology from the arXiv, "Deep Learning Machine Teaches Itself Chess in 72 Hours, Plays at International Master Level," MIT *Technology Review*, September 14, 2015.

4. John Markoff, *Machines of Loving Grace: The Quest for Common Ground between Humans and Robots* (New York: Ecco, 2016).

5. Carla Diana, "Don't Blame the Robots; Blame Us," *Popular Science*, December 2016.

6. Gary Marcus, *Rebooting AI: Building Artificial Intelligence We Can Trust* (New York: Vintage, 2019).

7. Kate Baggaley, "There Are Two Kinds of AI, and the Difference Is Important," *Popular Science*, February 23, 2017.

8. Chris Welch, "Google Just Gave a Stunning Demo of Assistant Making an Actual Phone Call," *The Verge*, May 8, 2018.

9. Marcus, *Rebooting AI.*

10. Ibid.

11. Susan Ratcliffe, ed., *Oxford Essential Quotations* (Oxford: Oxford University Press, 2016).

12. Carla Diana and Andrea Thomaz, "The Shape of Simon: Creative Design of a Humanoid Robot Shell." In *CHI EA '11: CHI '11 Extended Abstracts on Human Factors in Computing Systems* (New York: Association for Computing Machinery, 2011), 283–298.

13. Tom Guarriello and Carla Diana, "Dor Skuler, CEO and Co-Founder of Intuition Robotics," Episode 93, *The RoboPsych Podcast,* June 14, 2020. Podcast interview and follow-up email exchange with the author.

14. Michal Kosinski, David Stillwell, and Thore Graepel, "Private Traits and Attributes Are Predictable from Digital Records of Human Behavior," *Proceedings of the National Academy of Sciences of the United States of America* 110, no. 15 (April 9, 2013): 5802–5805.

15. Quantified Self website, October 20, 2020, https://quantifiedself.com.

16. Tom Guarriello and Carla Diana, "Brian Roemmele: The Last Interface," Episode 77, *The RoboPsych Podcast,* March 1, 2019.

17. Raffi Khatchadourian, "We Know How You Feel: Computers Are Learning to Read Emotion, and the Business World Can't Wait," *New Yorker,* January 19, 2015.

18. Affectiva website, October 20, 2020, https://www.affectiva.com.

19. Ibid.

20. Information Technology Research Centre, Telecommunication Research Institute of Ontario, "Ontario Telepresence Project: Final Report," March 1, 1995, chapter 2, 14–15, https://www.dgp.toronto.edu/tp/techdocs/Final_Report.pdf.

Chapter 9

1. Sarah O'Meara, "Coronavirus: Hospital Ward Staffed Entirely by Robots Opens in China," *New Scientist*, March 9, 2020; Amina Khan, "Meet Humanity's New Ally in the Coronavirus Fight: Robots," *Los Angeles Times*, April 11, 2020; Rachel Vigoda, "This Sushi Restaurant Takes Contactless Delivery to a New Level by Using a Robot," *Philadelphia Eater,* May 4, 2020.

2. Cade Metz and Erin Griffith, "A City Locks Down to Fight Coronavirus, but Robots Come and Go," *New York Times*, May 20, 2020.

INDEX

Note: Figures and tables are identified by *f* or *t* following the page number.

ABOUT THE AUTHOR

CARLA DIANA is a designer, author, and educator who explores the impact of future technologies through hands-on experiments in product design and tangible interaction. She has designed a range of products from robots to connected home appliances, and her designs have appeared on the covers of *Popular Science*, *Technology Review*, and *The New York Times Sunday Review*. She serves as Head of Design for Diligent Robotics, a company at the forefront of social robotics creating robots for health-care environments. She is an ongoing collaborator with the Socially Intelligent Machines Lab at the University of Texas at Austin, where advances in artificial intelligence and machine learning are manifest in expressive robots.

Carla writes and lectures frequently on the social impact of robotics and emerging technology and created the world's first children's book on 3D printing, *LEO the Maker Prince*. She cohosts the *RoboPsych* podcast, a biweekly discussion around design and the psychological impact of human-robot interaction.

Carla has taught and lectured extensively at universities throughout the world, and created some of the country's first courses in designing smart objects at the University of Pennsylvania, School of Visual Arts, and Parsons School of Design. In

2018 she was granted the honor of creating and launching the 4D Design program at the Cranbrook Academy of Art, where she serves as its first Designer in Residence.

Carla holds an MFA in 3D Design from Cranbrook Academy of art and a BE in Mechanical Engineering from the Cooper Union.